Horst Siebert (Ed.)

The World's
New Financial Landscape:
Challenges for Economic Policy

 Springer

Professor Horst Siebert
President
Kiel Institute of World Economics
D-24100 Kiel
http://www.uni-kiel.de/ifw/

Die Deutsche Bibliothek – CIP-Einheitsaufnahme

The world's new financial landscape : challenges for economic policy /
ed.: Horst Siebert. – Berlin ; Heidelberg ; New York ; Barcelona ; Hongkong ;
London ; Mailand ; Paris ; Singapur ; Tokio : Springer, 2001
     ISBN 3-540-41992-6

MK

Printed in Germany   SPIN   10836827

CONTENTS

N|A

B k title · **PREFACE**

ed.

The world's financial landscape has undergone profound changes in the past decades. "Globalization" has become a buzzword for economists, policy makers, and the press alike. Countries have increasingly opened up for foreign capital, the institutional structures of financial markets have been overhauled, and technology in the financial sector has advanced. A growing number of countries have been able to benefit from increased capital flows and investment, from access to new advanced technologies, and from the possibility to diversify risks. In Europe, monetary and financial integration has been stimulated by the recent introduction of the euro, and the benefits from financial integration are becoming visible. The flipside of these developments, however, is the greater exposure of economies to potentially volatile capital flows and to the risk of contagion, as recent crises in international financial markets have evidenced. The increasing integration of financial markets may imply new conditions for macroeconomic policies and eventually may require a redefinition of strategies that have worked well in the past. Now that the dust of recent financial crises has settled, it has become clear that there is crucial need for a better understanding of the factors driving international capital flows, of the changing structure of financial markets, and of the implications of globalized financial markets for economic policy.

Against this background, the Kiel Week Conference on The World's New Financial Landscape addressed the issue of finding the right concepts and institutional arrangements for the prevention and resolution of international financial crises. To better understand past events and to prepare for the challenges that will arise in the future, the conference participants studied the degree of international capital market integration that has been achieved over the past decades and analyzed the changes in the structure of financial markets that have accompanied the process of global capital market integration. The implications of the increased international mobility of capital and of the integration of emerging market economies into the world capital market for the regulation of financial markets, for the design of national policies, and for the institutional setup of financial markets were discussed from both a macroeconomic and a microeconomic perspective. In addition, the Kiel Week Conference offered an opportunity for a critical and in-depth evaluation of reform initiatives launched to reorganize the international financial landscape.

This volume contains the conference papers and the comments. In the first paper, *Claudia M. Buch* and *Christian Pierdzioch* provide a broad examination of concepts proposed in the literature to measure the integration of international

financial markets. They argue that a comprehensive picture of the degree of global capital mobility can best be obtained by combining different testing strategies. Studying the time trend of gross capital flows over gross domestic product,
they demonstrate that the sharp increase in capital flows observed over the second half of the 1990s reflects to a large extent growth in gross domestic product
per capita, increased trade links, deregulation of cross-border capital flows, and
increasing size of domestic financial systems. Furthermore, the authors report
that the structure of capital flows has undergone profound changes over time. The
general pattern that emerges is that securitized finance has become more important, especially for developed countries. Because, at the same time, the reliance on foreign direct investment tends to be negatively correlated with economic
development, the authors suggest that the pattern of capital flows to developed
and developing countries is consistent with the pecking order theory of international capital flows. In the second part of their paper, the authors point out that
empirical evidence suggests that the volatility of foreign direct investment might
lead changes in the volatility of portfolio and short-term other investment. All in
all, the complex links between different types of capital flows documented in
their paper indicate that there might be no optimal financial structure.

The complex interplay between, and the volatility of, global capital flows
might imply that the vulnerability of emerging market economies successively
integrating into international financial markets has increased over time. In the second paper, *William R. Cline* takes up this issue and provides a comprehensive
analysis of central aspects of financial crises management. An effective crisis
management allows a country to preserve the stability of its financial system, to
keep access to private international capital markets, and to return to economic
growth once the crisis has been resolved. Against this background, Cline carefully examines key factors that helped to resolve the crises in Mexico in 1995, Korea in 1997–98, and Brazil in 1998–99 and that caused Russia and Ecuador to
fail to manage successfully the crises that beleaguered these countries in the late
1990s. The special focus of the analysis lies on the potential role official international support and the private sector can play in the resolution of a financial market crisis. Studying the existing empirical evidence, the author emphasizes that
there is no clear evidence that private sector bailouts did entail excessive private
sector lending to emerging markets. He thus suggests that it might be the case
that, in recent crises, support by the international official community did not cause massive moral hazard risks. Resorting to the theory of sovereign debt to trace
out potential causes of a crisis, he argues that voluntary mechanisms for private
sector participation in crisis resolution should be established. Underpinning this
argument with a discussion of the experiences gathered in recent successfully
handled financial crises, he emphasizes that the adoption of such a crisis mana-

gement strategy allows a country to preserve reputation and, thus, access to the international capital markets.

Leslie Hull and *Linda L. Tesar* develop a theoretical framework that explains the structure of capital flows to developed and developing countries. They demonstrate that their analytical framework that features asymmetric information and relies on costly monitoring of loans by banks gives rise to a pecking order theory of international capital flows. The basic assumption on which their model is built has a long tradition in the corporate finance literature. It states that whether a company relies on bonds, bank loans, or equity as the primary source of external finance depends upon the relative ordering of the claims of bondholders and shareholders, upon the equity risk premium, and upon the benefits and costs of financial intermediation. The general prediction is that poorly rated companies resort to equity finance while firms with good credit ratings utilize the bond market to raise capital. Firms with an intermediate credit rating, in turn, rely either on bank loans or on equity. In terms of cross-border capital movements, the pecking order theory predicts that the relative importance of foreign direct investment (of securitized finance) as a source of external finance should decline (increase) when a country is on the transition path from a developing to a developed economy. Using their theoretical framework, the authors discuss the composition of external finance in a world populated by two identical economies, in a small open economy with firms with poor credit rating, and in a world characterized by international information asymmetries. They argue that the predictions of the theoretical model are in line with the patterns that can be detected in data on global capital flows.

Important determinants of the magnitude and structure of cross-border capital flows are the institutional features that characterize the international financial markets. The increasing integration of financial markets and the changing pattern of international financial flows have been accompanied by changes in the structure of the institutional setting of financial markets in general and in the role of banks in particular. *Gunter Dufey* provides a comprehensive discussion of these developments. He argues that the globalization of the financial services industry has opened up the opportunity for financial institutions to select between an increased number of competing regulatory regimes and that this development, in turn, has triggered the liberalization process that has taken place in world financial markets. Among the factors shaping the new landscape of the financial institutions sector are the advances made with respect to data collection and manipulation and data transmission capabilities. Technological progress has contributed to increasing the speed of internationalization and, given the arbitrage opportunities opened up by the existence of different national regulatory regimes, has implied that offshore markets have begun to play a more important role in the global financial system. This process, in turn, has put national regulatory systems under pressure. Dufey argues that these developments imply that economic policy makers will be forced to abandon regulations that were mainly established, for example, to shelter the domestic banking systems from competitive pressure. Those regulations that help to guarantee the efficiency and safety of the system, however, will survive. In particular, restrictions designed, for example, to enhance transparency and to force financial institutions to implement prudent risk management techniques will even gain in importance as the process of financial innovation is continuing to change the world's financial landscape. In this rapidly changing environment the core function of the banking system will still be to provide payment and transaction services.

The incentives for efficient investment and management decisions in public corporations rest crucially on the allocation of control rights, i.e., on the right to make decisions in situations not explicitly governed by written contracts. Empirical work on the allocation of corporate control has largely been limited to the United States because of data problems. Using a newly available data set, *Marco Becht* and *Colin Mayer* provide the most comprehensive study of corporate control structures in Europe yet. They find that in contrast to the United States and United Kingdom, control in continental European public corporations is typically highly concentrated. Moreover, the size distribution of controlling shareholdings suggests that controlling investors respond to the regulatory environment in maximizing private control benefits. Apart from regulation as a force shaping control structures, the authors also suggest that differences in control structures may be efficient responses to the technological characteristics of different firms and industries.

Paul De Grauwe and *Magdalena Polan* discuss the implications of increased international capital mobility for national macroeconomic policies. They start by observing that increases in international capital mobility make fixed exchange rate systems more fragile. Countries can respond to this development either by adopting a flexible exchange rate system or by establishing a monetary union. The authors stress that in a free-float regime countries face the problem of how to anchor nominal variables such as the price level. In addition, they discuss whether policy makers should be concerned about the high degree of exchange rate volatility typically characterizing freely floating exchange rates. All in all, the authors are rather skeptical about whether exchange rate target zones or currency boards can help to dampen exchange rate volatility because such regimes do not allow national monetary authorities to shelter an economy indefinitely from speculative attacks. They emphasize that one should not expect capital controls to help dampen the variability of real-world exchange rates. The reason is that capital controls not only discourage short-term speculation but also reduce the size of short-term hedging activities. The foreign exchange market is a multidealer market and if, for example, a transaction tax were implemented, the number of dealers might decline, which, in turn, would hamper the ability of the market to spread risks effectively. As regards monetary policy, the authors show that international capital mobility will increase the degree of monetary cooperation only if countries respond to this development by adopting more rigid exchange rate regimes. With respect to the implications of increased international capital mobility for the need for fiscal policy coordination, the authors focus on the situation in the European Monetary Union and stress that more intense budgetary policy coordination is required only if fiscal shocks trigger significant spillover effects in other countries. In addition, they discuss the role of fiscal coordination as an insurance mechanism against asymmetric shocks in a monetary union.

The contribution by *Tommaso Padoa-Schioppa* offers an in-depth discussion of the implications of increased capital mobility for the regulation of capital markets. He identifies the consequences of the free international movement of capital for the stability of the international financial system as the most important challenge for economic policy makers. He argues that such a close link between the free global movement of capital and the risk of financial fragility exists because supervision of the financial sector in many emerging market and transition economies is inadequate and because experience shows that financial markets are prone to contagion effects. Because returning to a system of financial repression would not constitute a solution to this problem but would instead eventually create additional problems, economic policy makers have begun to establish a market-led international monetary system. To this end, prudent risk management techniques have been designed, issues related to crisis prevention have gained in importance, and it has been acknowledged that the need for international policy cooperation has increased. Padoa-Schioppa discusses the characteristic features (concerning, for example, the allocation of responsibilities in the regulation of financial markets) of the institutional framework that has been designed to guarantee the stability of the international financial system. Taking this discussion as a point of departure, he goes on to discuss the four main challenges as regards the prevailing soft-mode system of financial market regulation: operationalizing the market-led soft-mode international monetary system by enhancing the functioning of international standard setters, enforcing regulatory rules in such a system, improving inter-agency cooperation, and dealing with offshore centers that open up the possibility of international regulatory arbitrage.

Focusing on the situation in emerging market economies, *Sebastian Edwards* constructs a new cross-country data set to shed light on the link between capital mobility and economic growth. Taking a broad perspective on various developed and developing countries, Edwards discusses historical periods characterized by downswings and upswings in the magnitude of international capital flows and, in particular, shows that capital flows tend to have been more volatile in emerging market economies. He then proceeds to compare alternative measures of the extent of international capital mobility. Building upon this comparison, he employs a new indicator of capital mobility to highlight that, with respect to economic performance, economies with a more open capital account appear to have outperformed countries that have restricted cross-border capital movements. He measures economic performance in terms of the growth of real gross domestic product and, alternatively, in terms of total factor productivity. A very interesting and highly policy-relevant insight offered by his analyses is that the link between economic performance and the openness of capital accounts seems to be nonlinear in the sense that the positive effect of an open capital account is positively correlated with the level of development of an economy. However, his empirical analysis also reveals that, at rather low levels of economic development, opening up the capital account might even exert a detrimental effect on economic performance. Thus, in this respect, emerging market economies might indeed be different from developed economies that have already established the financial institutions needed to manage cross-border capital movements effectively.

Philip R. Lane and *Gian Maria Milesi-Ferretti* present detailed empirical analyses of the determinants of the external capital structure of economies. Their point of departure is the observation that the level and the composition of international capital flows have changed substantially from bank lending to portfolio investment and foreign direct investment during the last decades. In a first step, they provide a comprehensive survey of the theoretical underpinnings of their empirical work. To this end, they survey important insights derived in the corporate finance literature on how asymmetric information, agency problems, and issues related to corporate control influence the optimal choice of firms between debt and equity. They discuss the applicability of their results to international macroeconomics and point out that, as compared with the settings typically examined in the corporate finance literature, additional problems related to the difference between domestic and foreign investors and issues regarding the enforceability of claims arise when cross-border capital movements are discussed. In a second step, they use these theoretical insights to conduct a broad empirical study on the determinants of the structure of international capital flows. Based on a huge data set on the stocks of external debt, direct investment liabilities, and portfolio equity liabilities of more than 100 developing and industrial countries, they estimate cross-sectional regressions to trace out important factors shaping the structure of the external capital structure of economies. In addition, the authors' findings will provide a basis for further discussions of differences in the external capital structure between industrial and developing countries and differences in the structure of capital flows between various regions of developing countries.

Dealing with a systemic liquidity crisis in the financial sector is the main task of a lender of last resort. In his contribution, *Frederic S. Mishkin* argues that such a lender of last resort not only can play a prominent role in a national context but can also help improve the functioning of the international financial system. To highlight the economics behind this proposition, Mishkin starts his analysis by discussing features that seem to be characteristic of a financial crisis. He argues that a financial crisis gathers steam when information asymmetries caused by adverse selection and moral hazard problems increase and, as a result, financial markets are no longer able to achieve an efficient allocation of funds. Focusing on the situation in emerging market economies, he argues that a financial crisis develops in two steps. In the first step, financial liberalization and a weak financial regulatory system result in a lending boom and excessive risk taking. In the second step, bank balance sheets deteriorate and investors realize that this diminishes the probability that monetary authorities will defend the domestic currency. Eventually, this chain of events might result in a currency crisis and, thus, a full-fledged financial crisis. Taking up this line of argumentation, he analyzes the role a lender of last resort can play in industrialized countries and emerging market economies. Based on the insights he derives, he develops eight fundamental principles that could help policy makers to guarantee that an international lender of last resort works effectively and to limit potential moral hazard problems arising when such an institution is established. He argues that the International Monetary Fund might be capable of serving as an international lender of last resort provided a reform of the Fund is agreed upon that provides this international institution with the instruments needed to manage financial crises in an effective way.

The Kiel Institute of World Economics is indebted to the participants in the 2000 Kiel Week Conference for having presented interesting and challenging papers and for having conducted stimulating discussions. The Institute gratefully acknowledges the financial support provided by the Deutsche Bank and by the Siemens AG. It would like to thank Commerzbank, Dresdner Bank, and Landesbank Schleswig-Holstein for their hospitality. It would also like to thank Claudia M. Buch, Hannelore Owe, and Christian Pierdzioch for helping to organize the conference, and Dietmar Gebert and Paul Kramer for preparing the conference volume for publication.

Kiel, May 2001   Horst Siebert

I.

Financial Market Integration:
The Track Record

Claudia M. Buch and Christian Pierdzioch

The Growth and Volatility of International Capital Flows: Reconciling the Evidence

1.   Motivation

"Globalization" has become a buzzword for economists, policymakers, and the press alike. Substantially increased capital flows, which have also become more volatile, may not only provide benefits in terms of a superior allocation of economic resources but also expose countries to the risk of abrupt reversals of foreign capital flows, thus forcing them into severe and prolonged real sector adjustment. In fact, the cumulative output losses of recent international financial crises have been quite substantial. Mussa et al. (1999) put them at values of up to 55–60 percent of GDP for Indonesia and Thailand, and even countries less severely affected such as Mexico or the Philippines reported losses on the order of 15 percent. These adverse effects of financial crises have given rise to a lively debate on policy issues, covering reforms of international financial institutions and the need to impose controls on certain types of capital flows. It goes without saying that the quality of any policy proposal depends on the quality of the underlying information set. What is needed is not only an assessment of the magnitude of international capital flows but also of their composition, their volatility, and of the potential interplay between various capital account items.

In view of the urgency of these policy issues, empirical research provides surprisingly little consistent and comprehensive information. As regards measures of the integration of international financial markets and thus the international mobility of capital, different concepts have been used, but each has its flaws and

Remark: The authors would like to thank Jörg Döpke, Stefan M. Golder, Ralph P. Heinrich, Jörn Kleinert, Lusine Lusinyan as well as the participants of the Kiel Week Conference for most helpful comments on an earlier draft. Jörg Breitung assisted in running ExploRe, the statistical software we used. Lusine Lusinyan and Marco Oestmann provided excellent research assistance. Remaining errors and inaccuracies are solely the authors' responsibility.

provides only limited answers to the questions at hand. As regards finding evidence of what determines the structure of international capital flows and its evolution over time, detailed research has only just begun.[1] As regards the volatility of capital flows, most research to date has focused on univariate measures of capital flows but has largely neglected the interplay of various capital flows, an exception being the work by Chuhan et al. (1996).

In view of the encompassing agenda for research on international capital flows, the aim of this paper can only be modest. Its main aim is to provide stylized facts on the evolution of global capital flows. Section 2 summarizes the evidence on international capital mobility and provides evidence on the determinants of capital flows for a panel of OECD countries for the post-Bretton Woods period. One point that this paper emphasizes is that there is no single comprehensive measure of capital mobility, as the standard tests consider individual market segments only or disregard characteristics of individual capital flows. Hence, at the end of this section, we go into more detail on the structure of capital flows, discussing, in particular, changes in the composition of capital flows. Having such information is of interest not only because different types of capital flows might have different implications for economic growth but also because much of the current policy debate centers around the issue that capital flows differ in their volatility characteristics. Section 3 looks at these issues in more detail. Our aim is to show whether the volatility of capital flows has increased over the past decades and which feedback effects between different types of capital flows we can identify. We provide new empirical evidence from German balance of payments data on these links. Our work confirms earlier research which finds that although FDI is less volatile than other capital flows, its volatility tends to precede changes in other capital account items. Section 4 concludes.

2.   Have Capital Markets Become More Integrated?

There is no generally agreed-upon measure of the degree of international capital mobility. Although various tests have been proposed in the literature, each has been criticized for providing only limited information on questions of interest or for being ambivalent in its interpretation. The problems start with the appropriate definition of capital mobility. At one extreme of the spectrum, capital mobility can be defined as the difference between national saving and investment; hence,

---

[1]   See Hull and Tesar (2000, 2001) or Lane and Milesi-Ferretti (1999, 2001). Buch (2000a) surveys evidence on the determinants of foreign assets of commercial banks.

capital would be considered mobile only if net foreign assets of a given country change during a certain period of time (Niehans 1994). The direction of change of net capital flows, in turn, is determined by differences in rates of return between countries.

At the other extreme, capital could be considered mobile if a large proportion of domestic investment is financed by foreign savings. If this measure is applied, large gross flows of capital relative to domestic investment needs would be taken as an indicator that capital is mobile internationally (Golub 1990).

In the following, we review briefly the different concepts that have been proposed in the literature to measure gross and net capital mobility. One problem that these tests have in common is the definition of a proper benchmark level, i.e., a measure of the degree of capital mobility under full integration. We use evidence both from the time of the Gold Standard, during which capital flows were essentially liberalized, and from regional financial market integration as such benchmarks.

More specifically, we start by reviewing evidence from interest parity or arbitrage tests, which focus on the hypothesis that interest rate differentials should induce (net) capital flows. Likewise, tests of the correlation between domestic savings and investment measure the responsiveness of net capital flows to differences in rates of return between countries. Portfolio tests, in contrast, essentially rely on the measurement of gross capital flows. Rather than considering relative rates of return only, these tests also consider risk characteristics and return correlations between domestic versus foreign financial assets. We conclude with an analysis of the evolution of gross capital flows relative to GDP for a panel of OECD countries and look at the changing structure of their capital flows.

a. Arbitrage Tests

In frictionless financial markets, identical financial assets must have the same price, irrespective of where they are being traded. This is the main rationale underlying arbitrage tests of financial market efficiency (or of capital mobility). The implementation of this definition is as difficult as the definition is simple because "identical" domestic and foreign financial assets must be identified, country and currency risk premia should be small, and transactions costs should be negligible.

Generally, there are three different kinds of interest parity: (i) uncovered interest parity, where differences in (nominal) rates of return on financial assets at home and abroad are equal to expected changes in the spot exchange rate, (ii) covered interest parity, where the forward discount replaces expected spot rates, and (iii) real interest parity, where uncovered interest parity and ex ante relative purchasing power parity obtain.

Real interest parity is thus a stronger condition than uncovered interest parity alone because it requires in addition that goods markets be perfectly integrated (Frankel 1989). By implication, real interest parity may fail to obtain even if uncovered interest parity holds when the (equilibrium) real exchange rate changes. This also implies that differences in real interest rates do not only reflect risk premia. Although empirical tests typically confirm nominal interest parity in the long run, they tend to find less support for real interest parity, mainly because purchasing power parity fails to explain (short-run) nominal exchange rate changes (Bayoumi 1999).

Even tests of nominal interest parity conditions pose substantial problems because — in a strict sense → they can be applied only to nominal interest rates for assets denominated in the same currency (Obstfeld 1995): $i_\$^* - i_\$ = 0$, where $i_\$ (i_\$^*) = $ the rate of return on domestic (foreign) assets or liabilities in US dollars. Comparing rates of return on assets in different currencies mixes restrictions on capital mobility with risk premia and expectational errors, all of which might drive a wedge between nominal interest rates without reflecting the immobility of capital. This can be shown by decomposing the ex post uncovered interest parity condition into

$$[1] \qquad i_\$^{US} - i_{DM}^G - \dot{e} = \left(i_\$^{US} - i_\$^E \right) + \left(i_\$^E - i_{DM}^E - \dot{e}^e \right) + \left(\dot{e}^e - \dot{e} \right) + \left(i_{DM}^E - i_{DM}^G \right),$$

where $i_\$^{US} (i_{DM}^G) = $ rates of return on local markets (United States and Germany) in local currency, $i_\$^E (i_{DM}^E) = $ rates of return on the euromarket, $\dot{e} = $ actual change in the exchange rate, and $\dot{e}^e = $ expected change in the exchange rate. Hence, the difference between the rate of return on domestic and foreign assets on each home market and the actual rate of change in the exchange rate equals the sum of (i) the on- and offshore interest differential for the foreign financial asset, (ii) the country risk premium (measured as the deviation from uncovered interest parity), (iii) an expectational error, and (iv) the on- and offshore interest differential for the domestic financial asset. Obstfeld therefore proposes focusing on off- and onshore deposit rates for the same currency (the first and the last terms on the RHS) in testing for interest parity because the term on the LHS in [1] incorporates information on a number of factors unrelated to capital mobility. Lacking information on the relative importance of these factors, we thus cannot conclude whether deviations from interest parity are due to capital account restrictions, asymmetries in information, expectational errors, or a combination of these factors.

The fact that interest parity tests can be applied in a meaningful way to *identical* financial assets only shows an important drawback in using such tests: they inform us about the degree of integration of very small segments of financial markets only. As soon as we consider market segments in which information costs matter, such as retail financial markets, interest parity tests should be applied only cautiously. Even for the United States, which is typically considered to be an internally integrated financial market, empirical studies tend to find a substantial dispersion of regional interest rates.[2] Since errors in the formation of exchange rate expectations are irrelevant in this context, evidence from the United States thus points to the importance of (implicit) regulatory barriers and, perhaps more importantly, information costs as a barrier to the full integration of financial markets.

A more fundamental drawback has been pointed out by Niehans (1994) who argues that arbitrage tests of financial integration provide basically no information about the (net) cross-border movement of capital. This is because prices of financial assets might change simply because they are bid up and down without corresponding changes in quantities. Niehans even concludes that "open economy macro-economics, to be reasonably realistic, should not be based on the arbitrage model of capital flows" (p. 28).

Despite these drawbacks, interest parity tests have remained the workhorse for testing the degree of international capital mobility. Taking a long-run perspective, Obstfeld and Taylor (1997) find that interest rate differentials between the United Kingdom and the United States were relatively small during the time of the pre-1914 Gold Standard. Subsequently, and in particular during the Second World War, interest differentials increased considerably; a similar increase could be observed in the late Bretton Woods period. Recent data show that there has been a decline in interest rate differentials to the levels observed at the turn of the century.[3] For Europe, Lemmen (1998) finds that covered interest parity conditions provide strong evidence that the degree of financial integration has increased over time. Likewise, Obstfeld (1995) shows that on- and offshore rates have been closely linked for most developed markets while arguing that less-developed countries have been less closely linked.

b.   Saving-Investment Correlations

Apart from arbitrage tests of capital mobility, the most often used and, at the same time, the most often criticized measure of capital mobility is the one suggested by Feldstein and Horioka (1980). They suggest looking at the correlation between domestic saving and investment by running the following regression:

---

[2]   See Buch (2000b) for a review of the evidence.

[3]   Similar results are obtained for a measure of real interest parity (Obstfeld and Taylor 1997).

[2] $\left(I / Y\right)_i = \alpha + \beta \left(S / Y\right)_i + \varepsilon_i,$

where I = domestic investment, S = domestic saving, and Y = gross domestic product of country i. Under perfect capital mobility, an increase in the saving rate in one country would cause an increase in investment in all countries. Estimates of β close to one could be taken as evidence for the incomplete mobility of capital. This measure of capital mobility implies a zero covariance between saving and investment and can thus be linked to the real interest parity condition. By decomposing the covariance between national saving and investment, it can be shown that real interest parity may, but need not necessarily, hold in order for the Feldstein–Horioka criterion of perfect capital mobility to be met (Lemmen 1998).

As regards the empirical measurement of β in equation [2], Feldstein and Horioka found a value of around 0.9. This result has been confirmed by a host of subsequent studies.[4] Although there is evidence for an increase in capital mobility in recent decades, international capital mobility would thus have remained imperfect. Taking a longer-term perspective, Taylor (1996) shows that the level of capital mobility that was approached in the early 1990s, with a β of 0.5 to 0.6, can be seen as a return to the levels observed already during the time of the Gold Standard.

Academic debate on the Feldstein–Horioka approach has addressed both the empirical methods and, more fundamentally, the concept of savings-investment correlations as such. As regards the empirical aspects, Krol (1996) has argued that the original approach, i.e., working with time-averaged data in cross-sectional regressions in order to eliminate business cycle effects, biases the results towards finding evidence for capital *im*mobility. He proposes working with annual data in panel regressions and controlling for business cycle effects by including a time dummy. Making these adjustments, Krol finds, in fact, lower beta coefficients, on the order of 0.2, for a panel of OECD countries, yet business cycle effects appear to be insignificant.

Hoffmann (1999), using the same dataset as Taylor (1996), reaches somewhat different conclusions. He argues that short- and long-run capital mobility must be distinguished and that the latter can be estimated using the Johansen cointegration procedure. Using this measure, long-run capital mobility for the United Kingdom and the United States appears to have been relatively high throughout the last century, the only interruption being the period of World War I. Variations

---

4   See Bayoumi (1990, 1999), Coakley et al. (1995), Montiel (1994), Sinn (1992), or Taylor (1996) for surveys.

in capital mobility shown by earlier studies would thus be due mainly to short-run changes in savings-retention coefficients.

Moreover, while Feldstein and Horioka used cross-section data, Gundlach and Sinn (1992) argue that these studies obscure different institutional structures between countries. They suggest exploiting the fact that the difference between saving and investment is the mirror image of the current account balance, i.e., of the difference between exports and imports of goods and services ($CA_t = X_t - M_t$), and thus propose testing for the stationarity of the latter: ·

$$[3] \qquad (I/Y)_t = \alpha + \beta (S/Y)_t + \varepsilon_i \Leftrightarrow (CA/Y)_t = -\alpha + (1-\beta)(S/Y)_t - \varepsilon_i.$$

Under the assumption that the error term ε is stationary, a nonstationary current account would imply that β is not equal to one, and that saving and investment move like independent random walks. Following the original interpretation of Feldstein and Horioka, this could be taken as evidence for capital mobility. Conversely, if the current account is stationary, β equals one, and capital is immobile. Gundlach and Sinn find that Germany, Japan, and the United States are integrated into the international capital market and find evidence for an increased degree of capital mobility in the post-Bretton Woods era.[5]

Following this line of reasoning, we performed tests on the stationarity of net capital flows relative to GDP for an unbalanced panel of 20 OECD countries for the years 1975 through 1998. These countries accounted for roughly 85 percent of global capital flows in the 1990s (Table 1).[6] We applied different unit root tests (Table 2). Levin and Lin (1993) adjust the standard ADF tests for unit roots to panel data, allowing for time trends and short-run dynamics. As in the ADF test, the null hypothesis that the variable contains a unit root is tested against the alternative that the variable is stationary. The IPS test proposed by Im et al. (1997) gives more flexibility with regard to the autocorrelation coefficient under the alternative by performing ADF tests for all cross sections and averaging over the estimated coefficients. The two tests in Table 2 yield different re-

---

[5] One problem with this approach is that the error term may not be I(0) if a country receives foreign aid. In this case, the current account may not be stationary even though private capital flows are low or even nil. Bagnai and Manzocchi (1996) argue that this problem can be solved by testing for stationarity of the current account minus the amount of aid that a country has received.

[6] Although this reveals the dominance of industrialized countries, it should not be overlooked, however, that developing countries and emerging economies have much larger shares in selected market segments, notably inflows of FDI.

Table 1: Structure of International Capital Flows in the 1990s (percent)[a]

	Share in global capital flows		Structure of capital flows	
	Industrialized countries[b]	Developing countries	Industrialized countries[b]	Developing countries
Assets	88.6	7.4	100.0	100.0
FDI	93.7	6.3	22.6	18.0
PI	93.7	4.5	38.1	21.6
OI	81.8	10.5	39.4	60.4
Liabilities	79.0	18.1	100.0	100.0
FDI	60.1	38.3	13.2	36.8
PI	84.7	13.0	46.2	31.0
OI	80.2	14.8	40.2	32.2
[a] Averages for the years 1991–1997. FDI = foreign direct investment; PI = portfolio investment; OI = other investments. — [b]Excluding international organizations.				

Source: IMF (1998).

sults for the entire sample period. While both agree that the first difference of net capital flows is I(0), the Levin–Lin test also indicates stationarity of the levels of net capital flows. The same holds true for two additional unit root tests.[7] Following the line of reasoning from above, this could be taken as evidence for the immobility of capital.

However, the approach of Feldstein and Horioka has been criticized on the grounds that the correlation of domestic savings and investment may be a statistical artifact if both are driven by omitted factors. These could comprise demographic factors, real interest rate developments, hysteresis of factor supplies, or government policies (Obstfeld 1995). Bayoumi (1990) in fact argues that government policy has been an important factor behind close postwar correlations of saving and investment, exerting its influence both directly through the imposition of capital controls and more indirectly through policies that targeted the current account.

An additional strand of the literature argues that a high correlation between domestic savings and investment may simply reflect the intertemporal budget constraint of an economy (Coakley and Kulasi 1997). The upshot is that a high (low) correlation of saving and investment in time series studies cannot *a priori* be taken as evidence of low (high) capital mobility. Hence, long-run correlations

---

7   See Breitung and Brüggemann (1999) for details.

Table 2: Panel Unit Root Tests[a]

	Levels		First differences		Degree of integration
	LL test	IPS test	LL test	IPS test	
	Capital flows (percent of GDP)				
Net capital flows	–1.75*	–1.28	–9.06*	–12.03*	?
Capital inflows	–1.33	–1.02	–6.86*	–10.58*	I(1)
Capital outflows	–0.57	–1.24	–7.57*	–11.39*	I(1)
FDI inflows	1.49	–0.71	–6.89*	–11.33*	I(1)
FDI outflows	1.87	–0.28	–5.59*	–9.15*	I(1)
OI inflows	–0.93	–1.13	–7.48*	–10.87*	I(1)
OI outflows	–1.78*	–3.06*	–7.58*	–12.79*	I(0)
PI inflows	0.36	–2.02*	–5.84*	–10.46*	?
PI outflows	4.58	1.56	–3.65*	–8.29*	I(1)
	Explanatory variables				
Log real GDP per capita	–2.80*	–1.75*	–3.11*	–3.63*	I(0)
Trade / GDP	–0.23	–0.89	–8.59*	–9.22*	I(1)
Log population	–2.89*	0.02	–0.56	–2.15*	?
Growth	–2.78*	–4.18*	–9.05*	–12.68*	I(0)
M2 / GDP	0.26	1.62	–3.98*	–5.94*	I(1)

[a] Specification with constant, trend, one lagged endogenous variable. * = significant at the 5 percent level. FDI = foreign direct investment; PI = portfolio investment; OI = other investments. The LL test is shown in Levin and Lin (1993), the IPS test in Im et al. (1997).

Source: Own calculations.

would indeed be expected to be close to one due to an intertemporal budget constraint, while low short-run correlations can be interpreted as indicators of capital mobility, a conclusion which has been confirmed by European data (Lemmen 1998).

Moreover, it could be argued that no useful benchmark exists for assessing the degree of international capital mobility based on estimates of β. Yet, as for interest parity tests, evidence for the degree of capital mobility at a national level can be used as such a benchmark. Generally, studies of capital mobility between different regions of a given country tend to find lower correlations between savings and investment than studies of capital mobility between countries (Bayoumi 1999; Bayoumi and Rose 1993; Helliwell and McKitrick 1998; Kellermann and Schlag 1999). The redistribution of savings through public transfers has been suggested as one.explanation (Bayoumi 1999). In view of the growing empirical and theoretical evidence for asymmetries in information between domestic and foreign investors,[8] information costs are also likely to prevent the free flow of capital across borders. At the same time, asymmetries in information are likely to affect the structure of (international) capital flows (Razin et al. 1998; Hull and Tesar 2000). Savings-investment tests, however, because of their aggregated nature, cannot reveal these links.

c.   **Portfolio Tests**

Capital market integration enhances economic welfare not only because it allows countries to draw on foreign savings to finance domestic investment but also because it provides the opportunity to optimally diversify investment portfolios. The more mobile capital is, the better these opportunities can be exploited. Portfolio tests of international capital mobility thus typically proceed by comparing an optimal, mean-variance efficient portfolio to actual portfolio choices. Although the optimal portfolio may, in principle, comprise all possible assets, including human capital (Lewis 1999), most empirical models look at securities portfolios only.

The standard model for analyzing international investment decisions is the capital asset pricing model (CAPM). This model is based on the assumption that there are no frictions in financial markets, that a risk-free asset exists, and that investors face no restrictions to sell short this riskless asset. It predicts that the risk of an individual financial asset is measured by its contribution to the overall variance of the expected excess return on the market portfolio over the risk-free rate, and that it will be priced accordingly.

According to the CAPM, the expected return on a financial asset, j, is given by the risk-free rate plus a risk premium which compensates investors for the risk relative to the market portfolio (m):

[5] $E(r_j) = r_f + [E(r_m) - r_f]\beta.$

<hr>

[8]   See Gehrig (1993), Gordon and Bovenberg (1996), or Razin et al. (1998) for theoretical models that assume differences in the information sets of domestic and foreign investors. Frankel and Schmukler (1996), Kim and Wei (1999), or Portes and Rey (1999) provide empirical evidence supporting this assumption.

This can be transformed into

$$[5']\quad E(r_j)= r_f + \left[\frac{E(r_M)-r_f}{\sigma(r_M)}\rho_{i,M}\right]\sigma(r_i),$$

where E = the expectations operator, r = the rate of return, σ = the standard deviation, r_f = the return on the risk-free asset, ρ = the coefficient of correlation, and β = the beta factor. For a positive (negative) correlation of the expected rates of return on the market portfolio and on asset j, the expected rate of return increases (decreases) with the standard deviation of asset j and with the expected return on the market portfolio. It decreases (increases) with the risk of the market portfolio. Hence, for a positive correlation, investors require a positive return compensation for higher risks. They are willing to accept greater standard deviations for asset j even at lower rates of returns if the correlation of asset j's return with that on the market portfolio is negative. This diversification of portfolios allows investors to insure against idiosyncratic risks. In an international context, such idiosyncratic risks are country-specific risks, whereas the nondiversifiable, systemic risk remains even in the world portfolio.

The main prediction of the CAPM is that investors hold combinations of the market portfolio and the riskless asset, the relative shares of which depend upon their degree of risk aversion. The optimal composition of the market portfolio, however, does not depend on investors' risk preferences. A test of the degree of capital mobility using the CAPM examines whether international investment portfolios are allocated according to these predictions. The overwhelming majority of the empirical papers finds that they are not. Rather, investors tend to hold the bulk of their financial assets in their home country and/or their home currency.[9] The causes of this home (or currency) bias are not well understood, although several explanations ranging from tax incentives, incomplete or asymmetric information, and the presence of nontradables to law enforcement problems have been advanced.[10]

---

[9] See French and Poterba (1990, 1991) or Tesar and Werner (1992). Kilka (1998) and Lapp (1999) provide evidence of this using German data.

[10] See Lewis (1999) or Obstfeld and Rogoff (1996).

d.   Gross Versus Net Capital Flows

Savings-investment correlations measure the magnitude of net rather than gross capital flows. Zero net capital flows might coincide with zero gross capital flows or with large capital in- and outflows which just so happen to be of a similar magnitude. This, in turn, would have completely different implications for the integration of financial markets. Hence, the focus on net capital flows unduly restricts the analysis of capital market integration. Also, it should be noted that an analysis of net capital flows is not conceptually identical to an analysis of real capital mobility. Just as zero net capital flows might coincide with large portfolio capital transactions, they might also be the result of large in- and outflows of FDI. Overall, then, gross capital flows can provide interesting insights into the links between national financial markets, not least because they reflect the degree of portfolio diversification going on.

Relating gross international capital flows to gross domestic financial flows, Golub (1990) reaches conclusions similar to those in the literature on net flows: capital mobility has been incomplete but on the rise. More specifically, Golub calculates the correlation between domestic asset issues and asset holdings. He argues that under complete capital mobility, the share of global financial assets held by a country should depend on the relative size of that country only. Comparing the share of a given country in total OECD financial wealth to the share of foreign to domestic assets and liabilities, he finds that these shares differ quite considerably for some countries. However, Golub's analysis has been criticized because it ignores the quite substantial differences in accounting methods used by various countries to value domestic and foreign assets (Bellak 1996). More conceptually, by simply comparing relative portfolio shares, insights provided by portfolio theory are ignored. These show that international portfolio decisions are based not only on relative market size but also on the interplay of risk, return, and return correlations.

Despite these criticisms, gross capital flows relative to GDP still provide a simple measure of financial market integration, although they need to be complemented by portfolio models in order to provide a benchmark for the degree of integration. Using the same panel as was used above to test for the stationarity of the current account, we thus calculated gross capital flows as the sum of foreign direct investment, portfolio investment, and other capital flows (mainly bank credits). Changes in the reserves and lending of international financial institutions were not considered.

Figure 1 shows the evolution of gross and net capital flows relative to GDP for 20 OECD countries since the early 1970s. A number of noteworthy features emerge from these figures:

First, gross capital in- and outflows have moved in a relatively parallel fashion for most countries.

Second, the magnitude of gross capital flows has been below 10 percent of GDP (in absolute terms) for most countries most of the time. Exceptions are countries hosting financial centers, such as Switzerland or the United Kingdom. More recently, countries like Ireland or the Netherlands have also reported gross capital flows on the order of up to 40 percent.

Third, while the share of capital flows to GDP was relatively stable in most countries under study initially, it grew quite rapidly in the second half of the 1990s. This was the case in countries like France, Germany, Italy, and the United States. Earlier studies of the degree of capital mobility covering the period until the mid-1990s might have failed to detect this recent development.

To gain insights into the determinants driving capital flows, we regressed both gross in- and outflows of foreign capital on a number of explanatory variables. We also broke total capital flows down into FDI, portfolio investments, and other investments although we were aware of the fact that standard balance of payments categories might not capture adequately the distinctions between these types of capital flows.[11] As explanatory variables, we used country size (log of population), the state of development (log of GDP per capita), and the degree of openness (volume of trade relative to GDP).[12] We would expect all of these variables to have a positive impact on the magnitude of gross capital flows and thus on the degree of integration into international capital markets.[13] Following Grilli and Milesi-Ferretti (1995) and Loungani et al. (2000), we constructed a dummy variable for the presence of capital controls. Using various issues of the IMF's Yearbook on *Exchange Arrangements and Exchange Restrictions*, we set the dummy equal to one if a country had restrictions on capital account transactions.

---

[11]   The distinction between FDI and portfolio investments is not always clear-cut. An investor might, for instance, buy equity shares in a foreign company. If this investment exceeded a certain threshold, the transaction would be registered as FDI, otherwise as portfolio investment. However, the nature of these investments in terms of the speed with which the investment can be reversed or the underlying movement of real capital might be very similar.

[12]   See Tables A1 and A2 in the Appendix for details on the data and the country sample. For similar applications of panel estimation techniques to capital flow data, see Rodrik and Velasco (1999), who analyze the determinants of short-term capital flows, or Calvo and Reinhart (1999), who assess the impact of sterilization policies and capital controls on capital flows over GDP for a set of emerging markets.

[13]   See Lane and Milesi-Ferretti (2001) for a more detailed discussion.

Figure 1: Gross and Net Capital Flows in Percent of GDP, 1970–1998

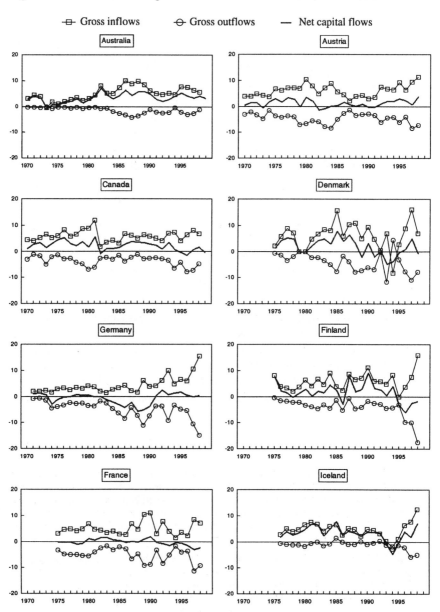

Figure 1: Continued

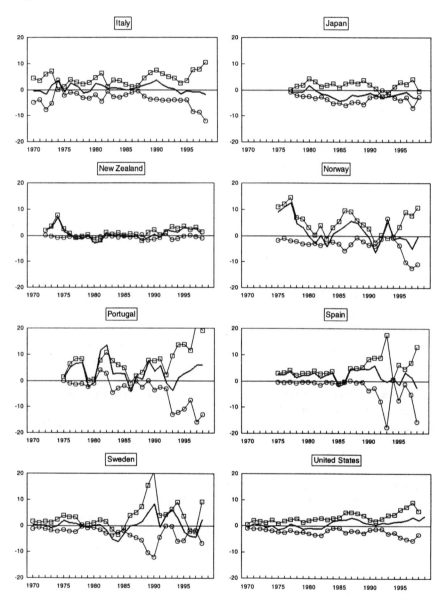

Figure 1: Continued

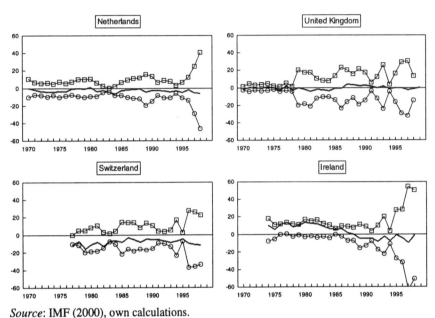

Source: IMF (2000), own calculations.

We started the analysis above by testing whether the time series under study were stationary; the results of the unit root tests are summarized in Table 2. Since, apart from one exception (other investment inflows), the hypothesis of nonstationarity was not rejected for capital flows, we used the two-stage Engle–Granger cointegration tests (Engle and Granger 1987).[14] For this purpose, the following equation was estimated in order to generate the long-run coefficients:

[6] $y_{it} = \alpha + \beta x_{it} + \varepsilon_{it}$,

where y_{it} = the ratio of capital flows over GDP in country i, x_{it} = time-varying explanatory variables, and ε_{it} = the error term. Then we tested, the residuals from estimating [6] for stationarity using the panel unit root tests described above.

The results from estimating [6] are summarized in Table 3.[15] Overall, we were able to explain roughly 25 percent of the variation in total capital outflows, out-

---

14   Also, three of the explanatory variables were found to be nonstationary.

15   In the final specification, data from only 17 countries were used because of insufficient observations for some time series for Iceland, Ireland, and New Zealand.

flows of FDI, and of portfolio capital outflows. The explanatory power of this equation is fairly low, in contrast, for capital inflows and for both in- and outflows of "other investments." The result for the latter category may be due to the fact that other investments comprise not only bank credits but also other capital account items, which may be driven by different factors. In the majority of the cases, there is evidence in favor of the hypothesis that the residuals are I(0), i.e., that the series are cointegrated. Evidence for cointegration is less clear-cut for FDI and portfolio investment outflows.

Notwithstanding these differences between the capital account items, some of our results are fairly robust across specifications: GDP per capita and the size of a country's financial sector tend to have a positive (and mostly significant) impact on gross capital flows. Likewise, country size measured by population also tends to have a positive impact, although the significance levels are somewhat lower. Trade openness has a significant impact on FDI flows relative to GDP only (supporting the hypothesis that the two are complements); the sign for portfolio investment even turns out to be negative (albeit insignificantly). In terms of economic significance, GDP per capita and population size dominate: a one percent increase in these variables raises total capital flows relative to GDP, at least proportionally. Elasticities with respect to trade and the size of the financial sector, in contrast, are much smaller.

Including dummy variables for EU membership and the presence of capital controls as reported in the bottom half of Table 3 leaves the baseline results largely unchanged. The capital control dummy has the expected sign in almost all equations and is significant for FDI and portfolio capital flows.[16] This result may seem at odds with the empirical literature, which finds a limited impact of capital controls on the volume and the structure of capital flows, in particular as regards their impact on the share of short-term capital flows. However, since we used a fairly broad measure of capital account restrictions and did not distinguish between controls on different forms of capital flows, our analysis provides only limited information on the ability of policy makers to affect particular capital account items by using administrative controls. Moreover, the abolition of capital controls has in many cases been paralleled by the deregulation of financial markets. Hence, we tend to interpret the capital control dummy as a fairly general measure of the deregulation trends in international financial markets that have taken place over the past three decades.

A positive and significant EU effect was found to obtain for foreign direct investment only, which would be in line with the argument that EU membership re-

---

[16] A dummy capturing different exchange rate regimes for capital and current account transactions, in contrast, is not significant.

Table 3: Determinants of Various Gross Capital Flows (percent of GDP)

	Total inflows	Total outflows	FDI inflows	FDI outflows	PI inflows	PI outflows	OI inflows	OI outflows
	Baseline regression[a]							
Trade/GDP	0.09**	0.16***?	0.05***	0.06***	−0.02	0.04*	0.05	0.05*
	(2.08)	(3.73)	(6.09)	(5.92)	(−0.62)	(1.82)	(1.47)	(1.68)
Log GDP per capita	5.10***	5.54***	1.50***	1.80***	2.21***	2.47***	1.38*	1.21
	(5.29)	(6.24)	(8.24)	(8.09)	(4.25)	(5.97)	(1.66)	(1.65)
Log popu-lation (−1)	1.32*	0.82	−0.08	−0.01	0.58	0.38	0.82	0.44
	(1.69)	(1.14)	(−0.53)	(−0.95)	(1.37)	(1.14)	(1.22)	(0.59)
M2/GDP	0.06**	0.06***	0.001	0.02***	0.03**	0.03***	0.03	0.01
	(2.38)	(2.69)	(0.10)	(2.91)	(2.43)	(3.01)	(1.21)	(0.56)
R² (overall)	0.03	0.24	0.04	0.32	0.01	0.18	0.02	0.09
	Residual unit root tests[b]							
LL (1993)	−4.65**	−4.59**	−1.95**	−2.95**	−2.65**	−2.60**	−3.55**	−3.15**
IPS (1997)	−2.24**	−1.32	−0.34	−0.31	−0.62	0.97	−1.78**	−3.13**
BB (1999)	−7.79**	−8.19**	−5.09**	−3.09**	−9.33**	−5.36**	−7.60**	−11.39**
Modified LL	−3.32**	−2.78**	−1.79**	−0.69**	−3.35**	−0.72	−2.55**	−4.08**
	Extended regression[a]							
Trade/GDP	0.07	0.09**	0.03***	0.04***	−0.02	0.01	0.06	0.03
	(1.52)	(2.01)	(3.21)	(3.95)	(−0.79)	(0.41)	(1.59)	(0.98)
Log GDP per capita	4.44***	3.23***	0.67***	1.18***	2.00***	1.53***	1.77*	0.46
	(3.63)	(2.93)	(3.04)	(4.27)	(3.04)	(2.98)	(1.68)	(0.49)
Log popu-lation (−1)	1.29**	0.78	−0.09	−0.02	0.55	0.34	0.82	0.45
	(1.65)	(1.10)	(−0.61)	(−0.13)	(1.31)	(1.05)	(1.22)	(0.75)
M2/GDP	0.06**	0.07***	0.002	0.02***	0.03**	0.03***	0.03	0.01
	(2.29)	(2.74)	(0.42)	(2.99)	(2.25)	(2.87)	(1.15)	(0.70)
EU	0.09	0.97*	0.46**	0.27*	−0.17	0.15	−0.21	0.56
	(0.13)	(1.69)	(4.10)	(1.88)	(−0.51)	(0.58)	(−0.37)	(1.17)
Control	−0.82	−1.59**	−0.36***	−0.41**	−0.66*	−1.13***	0.19	−0.05
	(−1.17)	(−2.51)	(−2.85)	(2.58)	(−1.75)	(−3.83)	(0.32)	(−0.09)
R² (overall)	0.03	0.32	0.12	0.36	0.02	0.24	0.03	0.12
	Residual unit root tests[b]							
LL (1993)	−4.35**	−3.51**	−3.77**	−3.64**	−2.47**	−2.53**	−3.64**	−2.98**
IPS (1997)	−2.43**	−1.48	−1.23	−0.73	−0.49	0.77	−1.69**	−3.25**
BB (1999)	−7.85**	−8.61**	−6.04**	−3.95**	−9.34**	−5.29**	−7.51**	−11.72**
Modified LL	−3.24**	−2.53**	−3.49**	−1.26	−3.56**	−0.59	−2.51**	−4.23**

[a]*** (**, *) = significant at the 1 (5, 10) percent level of significance. t-values in parentheses. Fixed effects estimates for the years 1975–1998. N = 369 (353) for estimates without (with) capital control dummy. — [b]** = the hypothesis that the residuals are nonstationary is rejected at the 5 percent level of significance. For details of the test specifications, see Breitung and Brüggemann (BB) (1999). LL stands for Levin and Lin (1993), IPS for Im et al. (1997).

Source: Own calculations.

duces uncertainty and thus particularly affects investments which show a high degree of irreversibility.[17]

As an additional explanatory variable, we included the growth rates of the economies under study.[18] Results were similar for the contemporaneous and the lagged growth rates: economic growth had a positive impact both an in- and out-flows of FDI and on inflows of other investments but not on portfolio investment or outflows of other investments. The remaining results, particularly the low explanatory power of the inflow equations remained largely unchanged.

It is interesting to note that our results partly support those of Lane and Milesi-Ferretti (2001). Although they look at the stocks of foreign liabilities of a much larger cross section of countries for 1997 only, whereas we looked at the flows for OECD countries, some results are similar. We found that openness for trade, the size of financial sector, and an increase in GDP per capita were associated with larger external liabilities. Likewise, Lane and Milesi-Ferretti found that capital controls had a more pronounced impact on equity (FDI) than on other capital flows. Also, links between trade and FDI were stronger than links between other capital flows and trade.

e.   The Changing Structure of International Capital Flows

Different market segments are likely to show different degrees of integration because institutional constraints and information costs differ. As these factors change over time, we would expect the structure of international capital flows to change along with them. Using aggregated measures of the degree of international capital mobility thus obscures important differences between market segments.

The literature on the determinants of the structure of international capital flows is still very much in its infancy. On the one hand, this is due to the limited understanding of the interplay between various forms of financing at the domestic level. On the other hand, explanations of international capital structures must additionally take account of aspects such as asymmetries in information between domestic and foreign investors, differences in institutional structures, or risks (foreign exchange, political) involved in international transactions. Since a review of this topic is the subject of other papers in this volume (Lane and Milesi-

---

[17]   For an application of the theory of investment under uncertainty to international capital flows see, e.g., Laban and Larrain (1997) or Bartolini and Drazen (1997).

[18]   Results are not reported but are available from the authors upon request.

Ferretti 2001; Hull and Tesar 2001), it suffices for the purpose of the present analysis to summarize the results of the so-called pecking order literature.

According to the original pecking order theory developed in a national context (Myers and Majluf 1984), firms tend to prefer internal finance. Should external finance become necessary, the safest securities are issued first, i.e., firms start by raising bank loans, then move to bond finance, and place new equity as a last resort only. Razin et al. (1998) develop a pecking order model of international capital flows by assuming differences in information between foreign investors and domestic managers. They argue that FDI has the advantage that it removes information barriers for foreign investors and that it allows access to superior management skills of the foreigners. As countries develop, information asymmetries are reduced and imports of technology become less important, and, hence, these advantages of FDI become less relevant.

In order to determine whether these predictions are supported by the data, we first compared the structure of international capital flows of OECD countries for the 1990s (Table 4).[19] According to the pecking order theory, we would expect international financial flows for this group of developed countries to be dominated by portfolio capital flows. Other investments (mostly bank loans) and FDI should be relatively unimportant. Yet, we found a clear dominance of securitized finance for both in- and outflows of capital for only three countries (France, Italy, Japan). Other investments, mainly bank loans and deposits, were dominant for yet a second group. While this could be attributed to a relatively low state of development for three of these countries (Portugal, Spain, Turkey), this was not the case for the other two (Switzerland, United Kingdom). No clear ranking was possible for the remaining countries.

However, when looking at developments over time, the trend towards an increasing disintermediation of international capital flows was borne out much more clearly. With one exception (Switzerland), the share of portfolio capital increased, while the share of other investments decreased for both capital in- and outflows. For Switzerland, in fact, the opposite trend was observed.

As regards the relative importance of FDI, in contrast, no clear time trend was visible. In some countries (Italy, Japan, Spain, Switzerland), the share remained constant, in some it increased (Austria, Denmark, France, Finland, Norway), and

---

[19]   We considered the post-Bretton Woods period only. This choice was dictated, first of all, by the lack of consistent data for earlier periods. In addition, the presence of a fixed exchange rate regime is likely to have affected the structure of capital flows during the Bretton Woods era. Hence, looking at longer-run trends would be an interesting issue for future research.

Table 4: Structure of Capital Flows to and from OECD Countries, 1970–1998
(percent of total)

	FDI inflows	Portfolio inflows	Other inflows	FDI outflows	Portfolio outflows	Other outflows
	Australia					
1970s	51.59	25.70	22.71	50.23	1.62	48.14
1980s	26.70	34.48	38.82	53.32	22.06	24.62
1990s	30.86	51.68	17.46	47.98	16.72	35.29
Total	30.37	43.40	26.23	50.18	18.26	31.56
	Austria					
1970s	5.94	23.04	71.01	3.06	6.80	90.14
1980s	6.01	44.53	49.46	6.45	18.31	75.24
1990s	16.16	56.70	27.15	15.66	48.70	35.64
Total	12.55	49.95	37.50	11.89	36.21	51.90
	Canada					
1970s	30.37	29.73	39.90	28.40	−0.84	72.44
1980s	17.78	53.45	28.77	38.75	15.11	46.14
1990s	29.16	54.63	16.21	43.77	36.70	19.52
Total	25.67	50.33	24.01	40.73	26.83	32.43
	Denmark					
1970s	...	...	...	...	...	...
1980s	3.84	11.06	85.10	14.85	12.40	72.75
1990s	31.92	50.21	17.87	35.24	33.08	31.68
Total	20.37	34.01	45.62	29.34	26.97	43.70
	Finland					
1970s	3.47	21.91	74.62	11.22	0.46	88.31
1980s	4.97	37.31	57.72	36.45	8.55	55.00
1990s	26.14	57.16	16.70	53.81	19.93	26.26
Total	17.71	48.49	33.80	48.14	16.46	35.40
	France					
1970s	10.01	6.65	83.35	7.57	4.44	87.99
1980s	11.10	25.71	63.19	19.85	8.72	71.42
1990s	25.14	42.23	32.63	33.40	48.30	18.29
Total	20.19	35.15	44.66	28.53	36.31	35.16
	Germany					
1970s	12.20	11.56	76.24	21.87	6.15	71.98
1980s	5.50	38.22	56.28	14.92	25.00	60.08
1990s	4.21	49.11	46.68	24.68	36.03	39.29
Total	4.90	45.12	49.98	21.98	31.48	46.54

Table 4: Continued

	FDI inflows	Portfolio inflows	Other inflows	FDI outflows	Portfolio outflows	Other outflows
Iceland						
1970s	...	...	...	...	...	...
1980s	...	...	...	...	...	...
1990s	13.78	53.32	32.90	21.12	65.73	13.15
Total	9.78	30.77	59.44	17.96	53.71	28.33
Ireland						
1970s	...	...	...	...	...	...
1980s	...	...	...	...	...	...
1990s	10.24	−0.87	90.62	2.99	6.13	90.89
Total	9.76	5.02	85.22	2.74	7.11	90.15
Italy						
1970s	9.17	−1.02	91.84	5.53	2.14	92.33
1980s	9.10	7.93	82.97	16.09	17.96	65.95
1990s	4.80	73.08	22.13	9.51	54.41	36.09
Total	5.91	55.77	38.32	10.15	46.62	43.23
Japan						
1970s	...	...	...	...	...	...
1980s	0.40	40.73	58.87	18.01	67.77	14.23
1990s	3.31	130.74	−34.05	20.31	62.04	17.64
Total	1.74	81.18	17.08	19.20	63.39	17.41
Netherlands						
1970s	14.78	11.44	73.79	34.27	3.84	61.90
1980s	20.11	25.89	53.99	33.63	15.15	51.23
1990s	27.56	29.01	43.43	35.87	38.97	25.16
Total	25.12	26.99	47.89	35.32	31.61	33.07
New Zealand						
1970s	...	...	...	...	...	...
1980s	...	...	...	...	...	...
1990s	64.30	11.38	24.33	39.33	43.97	16.70
Total	94.37	10.81	−5.18	59.66	28.06	12.29
Norway						
1970s	12.05	45.74	42.21	14.62	0.57	84.81
1980s	11.31	46.91	41.78	36.20	10.03	53.77
1990s	35.72	23.08	41.20	37.17	67.01	−4.18
Total	22.76	35.64	41.59	35.84	49.49	14.66

Table 4: Continued

	FDI inflows	Portfolio inflows	Other inflows	FDI outflows	Portfolio outflows	Other outflows
			Portugal			
1970s	9.45	-0.12	90.67	1.23	-0.31	99.09
1980s	31.36	32.85	35.79	7.80	0.50	91.70
1990s	15.57	27.22	57.22	10.23	33.44	56.33
Total	17.21	26.97	55.83	10.01	31.77	58.22
			Spain			
1970s	21.17	0.16	78.68	18.28	1.31	80.41
1980s	43.29	21.71	35.01	42.65	9.70	47.65
1990s	21.94	33.31	44.74	16.70	25.56	57.75
Total	25.28	30.11	44.61	17.72	24.66	57.61
			Sweden			
1970s	6.02	17.20	76.77	44.92	0.27	54.82
1980s	12.26	6.77	80.97	60.37	10.32	29.31
1990s	58.47	−64.79	106.33	90.33	−45.30	54.98
Total	40.89	−37.52	96.63	76.83	−22.55	45.72
			Switzerland			
1970s	...	...	...	...	...	...
1980s	8.37	28.56	63.06	12.25	39.47	48.28
1990s	10.63	18.73	70.64	20.35	29.62	50.03
Total	9.80	21.09	69.11	17.05	33.45	49.50
			Turkey			
1970s	...	...	...	...	...	...
1980s	...	...	...	...	...	...
1990s	14.23	20.66	65.11	6.17	33.74	60.09
Total	14.20	20.89	64.92	5.12	28.19	66.69
			United Kingdom			
1970s	23.08	7.53	69.39	31.74	2.42	65.84
1980s	11.46	20.06	68.48	19.46	20.19	60.35
1990s	15.22	26.47	58.31	28.23	27.72	44.05
Total	14.45	23.73	61.82	25.81	24.32	49.87
			United States			
1970s	10.63	33.78	55.59	33.24	10.22	56.54
1980s	21.72	32.09	46.19	21.12	8.52	70.35
1990s	22.82	49.04	28.14	35.78	34.12	30.10
Total	21.91	43.87	34.22	32.00	25.74	42.26

Source: IMF (2000), own calculations.

in some there were opposing trends for in- and outflows.[20] The immediate explanation would be that FDI flows were driven by a number of factors unaccounted for in the pecking order models.

The question remains as to what extent trends in the structure of financial flows of OECD countries inform us about global trends in capital flows. Edwards (2001) and Lane and Milesi-Ferretti (2001) find, for instance, that characteristics of capital flows to developed countries and emerging markets differ significantly. Also, the theoretical models of international capital flows eluded to above provide strong support for the hypothesis that the structure of international capital flows changes over the course of economic development.

Mussa et al. (1999) analyze the pattern of capital flows to emerging markets. They find pronounced secular swings in net capital flows with the financial crises of the 1980s and the late 1990s being preceded by episodes of surges in capital inflows. However, the structure of foreign capital during these inflow periods differed quite substantially: prior to the 1980s, capital flows were dominated by flows of funds to the public sector, whereas private-to-private capital flows dominated later on. Also, portfolio investment increased relative to bank loans, thus paralleling the trend observed for developed markets.

Yet, differences between developed and emerging markets were also evident in the data (Table 1). In the 1990s, capital inflows to industrialized countries were dominated by portfolio investments and other investments. FDI accounted for a relatively small share of 13 percent. Developing countries, in contrast, relied on FDI much more heavily. In fact, the three sources of finance were of roughly equal importance. As for capital outflows, the pattern for industrialized countries by and large resembles the pattern for inflows. In the case of developing countries, in contrast, other investment outflows dominated (60 percent), and FDI and portfolio investment were of similar importance.

Differences in the structure of capital flows between the two groups of countries are also reported by Hull and Tesar (2001). For industrialized countries, the observations that capital in- and outflows are of a similar magnitude and tend to be reinvested in these countries support the view that portfolio diversification appears to be the main motive for international capital flows. Reflecting the state of development of these economies, bonds are more important than bank loans and equity. Moreover, developing countries with low credit ratings appear to move up the pecking order of international finance as they open up for foreign capital. For these countries, the composition of capital flows tend to be skewed towards equity finance and bank loans. Finally, equity finance tends to dominate in the

---

[20]   Japanese capital outflows and Norwegian capital inflows being the sole exceptions.

case of developed markets, indicating that information asymmetries are important.

An additional piece of evidence which is of interest when analyzing trends in
international capital flows is the share of short-term capital. After all, a high exposure to, and large swings in, short-term capital are often held responsible for
the occurrence of currency crises. Unfortunately, however, standard balance of
payments statistics give only insufficient account of this share. Inflows of financial credits and portfolio capital flows are often not classified according to
their maturity. Data provided by the Bank for International Settlements (BIS) on
the maturity structure of bank lending to countries outside the BIS reporting area
can be used as an indicator though (Figure 2). These data show that the share of
short-term loans in total lending shifted upward when comparing the 1990s to the
1980s. While, roughly 40 percent of all foreign loans had a maturity of less than
one year in the 1980s, this share increased to more than 50 percent in the 1990s.
Recently, it has come down again.

Overall, Mussa et al. (1999) note that there does not seem to have been a
secular trend towards an increasing share of short-term foreign debt in recent
decades. Yet, they confirm the evidence presented in Figure 2: remaining maturi-

Figure 2: Short-Term Loans as Percent of Total Foreign Loans Granted to
Countries Outside the BIS Reporting Area, 1980–1998

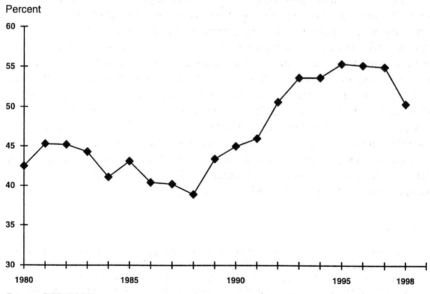

Source: BIS (1999).

ties tended to become shorter during the boom phase of large capital flows between the late 1980s and mid-1990s, while they became longer afterwards.

Empirical work shows that the share of short-term loans is a positive function of GDP per capita and the size of the financial sector of the host country (Rodrik and Velasco 1999; Buch and Lusinyan 2000). Generally, the results of such work suggest that determinants of short-term bank loans are similar for developed and developing countries. Moreover, regulatory restrictions have an impact on the maturity structure of foreign bank lending. OECD membership, in particular, has a negative impact on the share of short-term foreign loans, due to the risk-weighting implied in the BIS capital adequacy standards.

f.   What Have We Learned?

On a most general level, this section has shown that there is no universal measure of the degree of capital mobility. Tests focusing on the magnitude of net capital flows assess the effects of rate of return differentials on capital mobility. Tests focusing on the magnitude of gross capital flows typically take additional account of risks and return correlations. According to these measures, a high degree of capital mobility might coincide with large or small net capital flows. A comprehensive assessment of capital mobility should thus consider both types of tests.

An additional distinction has been made between price and quantity measures of capital mobility. Price or arbitrage tests make use of the fact that in integrated financial markets rates of return on identical financial assets must be the same. Alternatively, savings-investment tests have been used because, in integrated financial markets, domestic investment should not be constrained by the supply of domestic savings.[21] It is a relatively common finding in the empirical literature that price measures show a greater degree of integration than quantity measures (Bayoumi 1999). One explanation for this dichotomy is that two different time horizons are applied. While interest parity tests are typically based on return data for relatively short-term financial assets, quantity tests look at the intertemporal allocation of capital and are of a decidedly more long-term nature. In addition, arbitrage tests are restricted to relatively small market segments only, whereas savings-investment tests are based on total capital flows.

---

[21]   Alternatively, tests of the correlation between consumption and net domestic output can be used to assess the degree of capital mobility (Shibata and Shintani 1998). Like tests of savings-investment correlations, these tend to show lower degrees of capital mobility than interest parity conditions. However, they do not show a clear trend towards integration for developed countries (Obstfeld 1995).

In addition, changes in the volume of capital flows relative to GDP provide information about the degree of integration into international capital markets. Panel estimates for OECD countries have revealed that GDP per capita, openness for trade, the size of the financial sector, and financial deregulation have had a positive influence on the degree of financial market integration. For developed countries at least, legal factors in the form of outright restrictions to the free flow of capital thus seem to play a negligible role in limiting the free flow of capital across borders.

For all measures, however, the *degree* of capital mobility can be assessed only by defining an appropriate benchmark level. In the literature, two such benchmarks have been used. The period of the Gold Standard is typically considered as a period during which legal capital account restrictions were minor. When comparing the degree of capital mobility today to that prevailing during that time, both interest parity and savings-investment tests produce the same result: capital mobility may be higher today than it used to be a couple of decades ago, but not necessarily higher than at the turn of the last century.[22]

A second benchmark is the degree of integration of financial markets at a national level. Most tests of capital mobility indicate that the degree of international capital mobility is lower than the degree of integration of national financial markets. These discrepancies suggest that institutional differences and information costs effectively separate national financial markets of even those economies which have attained a relatively high degree of financial openness in legal terms. Institutional factors and information costs, however, are captured only insufficiently by standard tests of capital mobility for two reasons:

First, tests of the degree of capital mobility typically provide little information on the degree of financial integration defined as the links between domestic and international financial markets (von Furstenberg 1998). That is, they do not show to what extent domestic financial markets are actually exposed to competitive pressure from abroad and to what extent implicit institutional barriers to market entry prevail. While treatment of this issue is certainly beyond the scope of the present paper, the example of Germany shows that the two aspects are not necessarily linked. While the ratio of gross capital in- and outflows to GDP shows an increasing degree of integration of Germany into international capital flows (Figure 1), the penetration of the domestic banking market by foreign banks remains fairly low. Foreign financial institutions account for a relatively stable 3–5

---

[22] Unfortunately, no evidence exists concerning changes in the importance of the home bias in investment portfolios over time.

percent of retail banking activities (Buch and Golder 2000). Similar observations can be made for other markets in the EU (ECB 1999).

Second, most tests of the degree of capital mobility disregard the sensitivity of financial market segments to information costs. Because of the substantial advances that have been made in information technology and the reduction in the degree of market segmentation due to asymmetries in information, it might thus be premature to conclude that the nature of capital market integration then and now has been the same. Evidence collected by Bordo et al. (1998) in fact shows that the type of investments financed on international capital markets has shifted away from large-scale infrastructure projects conducted by large public borrowers towards less tangible projects conducted by private entrepreneurs.

As information on small and mid-sized borrowers tends to be more difficult to obtain, this change in the structure of international finance could be taken as evidence of a greater availability of information. Since, however, we have very limited means of testing explicitly for the severity of information problems, one should be careful not to draw quick conclusions. It is well conceivable that investors in the late 19th century were as poorly informed about the prospects of large-scale (foreign) infrastructure projects as are investors today who need to assess the feasibility of an investment project of, say, an internet startup. Incidentally, if incomplete information sets are a major factor in incomplete mobility of capital, the fact that aggregate measures of capital mobility show that the degree of integration of financial markets one hundred years ago was similar to the degree nowadays might be taken as an indication that, in relative terms, information has not become more readily available.

3.   What Do We Know about the Volatility of Capital Flows?

Recent financial crises have shifted interest away from measures of the degree of capital mobility towards measures of the volatility characteristics of different capital account items. In the debate on the risks and benefits of capital account liberalization, it has been pointed out that capital flows to emerging markets often tend to be relatively volatile and that this may hold in particular for short-term capital flows.[23] Consequently, there has been a lively policy debate on possible measures to shield emerging markets from volatile capital flows.

---

[23]   See Buch (1999) for a review of the literature.

In this section, we take up this debate and analyze the volatility of cross-border capital movements. First, we survey the (surprisingly scarce) existing empirical evidence on capital flow volatility. Second, we augment the existing evidence by taking a closer look at the dynamic relation between the volatilities of different capital account items for Germany. Though restrictions on the availability of comprehensive balance of payments data preclude an empirical analysis of a broader sample of countries, the analyses reveal interesting additional insights into the interplay between the volatility of, for example, foreign direct and portfolio investment flows and suggest potentially fruitful avenues for future research. Third, we complete our analysis by focusing attention on the frequency at which large and substantial capital account reversals have occurred over the past three decades.

a.   Earlier Empirical Evidence

In contrast to the extensive policy debate on volatile capital flows, empirical evidence on volatility patterns is rather scarce. Claessens et al. (1993) have been among the first to draw attention to the fact that standard balance of payments categories provide relatively little information on the actual volatility of capital flows. Classifying balance of payments data for five industrial and five emerging economies by type of instrument and type of transactor, Claessens et al. use quarterly changes of capital flows for the years 1973–1992. The persistence of capital flows is first assessed by looking at the autocorrelation features of capital flows. Additionally, they calculate half-lives using impulse-response functions and the Q-statistics on the joint significance of the first 16 autocorrelations. One result of their analysis is that there is no consistent pattern of persistence across different capital flows. Also, short-term capital flows tend to be at least as predictable as other capital flows. Their results also show that long-term capital flows tend to be the most accommodating[24] and that there is substantial substitution between different types of capital flows.

An alternative quantitative method to recover differences in the magnitude of the permanent and transitory components of capital flows has been suggested by Sarno and Taylor (1999). Resorting to a Kalman filter technique and variance ratio tests, they estimate the unobserved permanent and transitory components contributing to capital flows to a sample of nine Latin American and Asian developing economies for the years 1988–1997. Their results suggest that both private portfolio investments (comprising equity and debt flows) and official flows

---

[24] That is, they respond most sharply to changes in the overall capital account.

are driven by a very strong temporary component. Changes in the permanent component, in contrast, serve to explain a large fraction of the volatility of cross-border commercial bank credits and foreign direct investment flows.

In contrast to this paper, the focus of a recent paper by Chuhan et al. (1996) has been on the interplay between alternative categories of capital flows. Their central argument is that univariate studies might unearth spurious similarities among alternative forms of capital flows whenever one category of cross-border capital flows can be identified as an important factor driving other components of the capital account. They use quarterly data for several developed and emerging economies for the years 1985–1994. Bivariate tests for Granger noncausality for alternative pairs of capital flows indicate that (changes in) FDI might account for a substantial fraction of the subsequent variation in short-term investment.

This result is further confirmed by the outcomes of variance decompositions which Chuhan et al. carried out for multivariate VARs comprising all capital flows for each country. For most of the countries, the historical volatility of short-term capital flows can largely be explained by the orthogonalized shocks entering into the equation for FDI. The reverse does not hold. A further interesting result is that short-term investment flows to one country tend to be more sensitive to changes affecting short-term investment flows to other countries, whereas, in the majority of cases, similar conclusions cannot be drawn when FDI flows are considered. Thus, bilateral cross-country evidence indicates that disturbances in international financial markets might spill over more easily onto short-term investment flows than onto foreign direct investment flows.

Further evidence of the volatility of capital flows has been documented by Mussa et al. (1999). Using gross private capital flows to emerging markets, they find that the variance in capital flows was higher in the second half of the 1990s than in earlier periods. However, they do not find such a general pattern when net capital flows are used. When they break down total capital flows into their components, FDI turns out to be the most stable, followed by portfolio capital and bank credits.

We obtained similar results for the capital flows to developed countries. Table 5 shows the evolution of the coefficients of variation of net foreign direct investment (FDI), net portfolio investment (PI), and net other investment flows (OI) for 13 developed countries over the last three decades.[25] The balance of payments data were sampled at a quarterly frequency for the period 1971:1–1998:4. They were taken from the *International Financial Statistics* of the IMF. With the exception of Portugal and Sweden, the time series of "other in-

---

[25]   We restricted the sample used above to those countries for which long-run time series without large outliers were available.

Table 5: Coefficients of Variation of Net Capital Flows, 1970–1998[a]

Country	Type of capital flow	Coefficient of variation		
		1970:1–79:4	1980:1–89:4	1990:1–98:4
Australia	FDI	0.48	0.87	0.71
	PI	1.84	1.01	0.94
	OI	1.29	0.73	2.23
Austria	FDI	0.46	0.90	0.78
	PI	0.85	0.67	0.66
	OI	1.77	2.33	2.64
Canada	FDI	0.47	0.71	0.65
	PI	1.01	0.67	0.64
	OI	0.79	1.94	2.63
France	FDI	0.41	0.96	0.33
	PI	0.46	0.91	0.96
	OI	1.19	1.29	4.93
Germany	FDI	0.40	0.78	1.11
	PI	1.22	0.93	0.64
	OI	1.06	1.01	0.97
Italy	FDI	0.78	1.00	0.47
	PI	183.17	2.07	1.01
	OI	2.73	2.16	1.34
Japan	FDI	...	1.00	0.36
	PI	...	0.89	0.93
	OI	...	1.30	...
Netherlands	FDI	0.50	0.74	0.84
	PI	0.78	0.94	1.01
	OI	1.24	1.41	2.64
Portugal	FDI	0.72	1.15	0.66
	PI	11.27	2.52	1.15
	OI	0.90	2.41	1.16
Spain	FDI	0.44	0.75	0.47
	PI	28.62	1.61	1.29
	OI	0.35	1.75	1.32
Sweden	FDI	0.39	0.95	1.10
	PI	0.99	3.06	1.59
	OI	0.90	1.70	1.71
UK	FDI	0.59	0.78	0.90
	PI	1.59	1.10	0.89
	OI	2.38	0.82	1.43
USA	FDI	0.56	0.66	0.63
	PI	0.77	0.80	0.64
	OI	1.03	0.87	1.33

[a] For France, Portugal, Sweden, and Spain data were available for 1975–1998 only. Coefficients of variation were computed by dividing the standard deviation of the respective time series by the absolute value of their mean value. FDI = foreign direct investment; PI = portfolio investment; OI = other investment.

Source: IMF (2000), own calculations.

vestment" was identified as the most volatile component of the capital account. Moreover, for most countries, the coefficients of variation of other investment increased over time. The exceptions are Germany, Italy, Portugal, and Spain. For Germany and Italy, the reported coefficients of variation of other investment even declined over the entire sample period. For Portugal and Spain, the variability of other investment reached a maximum in the 1980s. The evidence reported for the United Kingdom and the United States, in contrast, suggests that the volatility of other investments was lower in the 1980s than in the preceding decade but increased again during the 1990s.

As regards the dynamics of the volatility of portfolio investment flows, seven out of the thirteen countries under investigation experienced a monotonous decline in the volatility of this item of the capital account. Furthermore, while foreign direct investment seems to be the most stable component of capital flows, the coefficients of variation show neither a general tendency to increase nor to decrease.

In order to provide further insights into the evolution of the variability of cross-border capital movements over time, Table 6 shows the corresponding coefficients of variation computed for gross capital flows. Visual inspection of the table indicates that outflows of capital categorized as "other investment" tend to be more volatile than foreign direct and portfolio investment outflows. In the case of capital inflows, a similar proposition does not hold for Germany, Portugal, Spain, the United Kingdom, and the United States. Moreover, the table suggests that the variability of other investment flows was somewhat higher in the 1990s than in the 1980s. As in the case of net capital flows, no clear-cut tendency towards increase or decrease can be detected with respect to the volatility of gross foreign direct and gross portfolio investment flows. On a comparative basis, however, foreign direct investment can be said to be the most stable component of the capital account. Thus, by and large, the figures presented in Table 6 serve to corroborate the impression already obtained by analyzing the dynamics of the coefficients of variation of net capital flows.

It goes without saying that the empirical evidence presented in Table 5 and Table 6 can be criticized on the ground that the sample of countries included in the analysis consists of mature economies only. There is evidence that volatility of capital flows to developing countries exceeds that of developed economies (Edwards 2001), although general time trends seem to be similar: While overall volatility of capital flows increased in the 1990s, this increase was mainly driven by an increase in the volatility of debt flows. Volatility of FDI and portfolio capital flows seems to have declined, in contrast. Nevertheless, the tables suggest that, at least for the developed countries in the sample, volatilities of capital flows are comparable up to a certain degree.

Table 6: Coefficients of Variation of Gross Capital Flows, 1970–1998[a]

Country	Type of capital flow	Inflows			Outflows		
		1970:1–79:4	1980:1–89:4	1990:1–98:4	1970:1–79:4	1980:1–89:4	1990:1–98:4
Australia	FDI	0.55	0.98	0.81	0.48	1.20	1.01
	PI	1.77	0.96	0.87	3.99	1.88	3.35
	OI	1.27	0.74	2.66	2.26	2.01	2.77
Austria	FDI	0.46	0.83	1.06	0.79	1.16	0.51
	PI	1.03	0.74	0.70	1.25	0.97	1.07
	OI	1.95	2.44	2.78	1.65	2.37	2.76
Canada	FDI	0.35	1.14	0.68	0.83	0.53	0.72
	PI	0.99	0.69	0.86	7.15	1.22	0.62
	OI	0.89	2.03	3.29	0.86	2.15	2.90
France	FDI	0.47	0.96	0.39	0.47	1.01	0.44
	PI	0.71	1.02	1.56	0.88	1.15	1.00
	OI	1.27	1.33	4.15	1.14	1.30	6.88
Germany	FDI	0.50	1.96	3.19	0.54	0.69	0.93
	PI	1.50	1.53	0.67	2.09	1.09	0.92
	OI	1.33	1.21	0.93	1.10	1.11	1.45
Italy	FDI	0.77	1.50	0.62	1.27	0.74	0.57
	PI	3.77	3.52	0.90	6.57	1.61	1.58
	OI	2.47	1.66	3.51	3.15	3.36	1.44
Japan	FDI	...	4.39	1.43	...	1.04	0.35
	PI	...	1.72	1.53	...	0.88	0.89
	OI	...	1.58	...	...	1.14	...
Netherlands	FDI	0.61	1.00	1.10	0.54	0.67	0.73
	PI	1.24	1.23	1.42	1.85	1.20	0.97
	OI	1.34	1.51	2.09	1.19	1.44	3.71
Portugal	FDI	0.64	1.16	0.64	4.66	1.91	1.21
	PI	14.14	1.73	1.63	1.37	7.49	1.18
	OI	1.10	3.25	0.98	1.08	4.29	1.82
Spain	FDI	0.53	0.75	0.39	0.48	0.93	0.96
	PI	15.31	1.76	1.90	2.56	3.11	1.70
	OI	0.44	2.01	1.24	0.64	3.35	1.82
Sweden	FDI	0.78	0.97	1.19	0.36	0.98	1.14
	PI	1.37	4.83	1.69	32.87	2.35	1.60
	OI	0.96	2.03	1.70	1.29	1.13	2.60
UK	FDI	0.59	1.01	0.76	0.70	0.75	1.09
	PI	1.86	1.60	1.24	3.60	1.20	1.21
	OI	2.41	0.87	1.26	2.36	0.85	1.77
USA	FDI	0.61	0.73	1.09	0.62	0.83	0.46
	PI	0.93	0.85	0.74	0.73	1.11	0.71
	OI	1.20	0.88	1.14	0.96	1.03	1.94

[a] For France, Portugal, Sweden, and Spain data were available for 1975–1998 only. Coefficients of variation were computed by dividing the standard deviation of the respective time series by the absolute value of their mean value. FDI = foreign direct investment; PI = portfolio investment; OI = other investment.

Source: IMF (2000), own calculations.

A second drawback of the evidence presented in Table 5 and Table 6 is that the capital flows of the category "other investment" comprise both short-term and long-term cross-border capital movements. The tables suggest only that other investments are the most volatile capital flows. They do not take the relative volatility of short-term versus long-term cross-border capital movements into account. Furthermore, the tables give an impression of univariate properties of the variability of the different components of the capital account only. Therefore, it is not possible to draw inferences regarding the interplay of the alternative categories of capital flows. The latter two arguments will be explored next.

b.   Links between the Volatility of Capital Flows: The Case of Germany

Earlier empirical studies on the volatility of capital flows have largely confirmed the conventional wisdom that short-term capital is more volatile than long-term capital, in particular FDI. Yet, the empirical evidence presented in Chuhan et al. (1996) suggests that not only the univariate properties but also the various relations between different forms of capital flows may provide interesting and important insights into the dynamics of the capital account. In particular, their results indicate that FDI flows are more stable than short-term capital flows but, at the same time, that they are also one of the driving forces behind the dynamics of other items of the capital account. Hence, this section provides more evidence on the volatility features of particular capital account items and the interplay between them.

Since we cannot use annual data for such an analysis and since quarterly or even monthly capital account data are not generally available for a larger set of countries, we use German data only. Though this implies that we cannot provide potentially fruitful cross-country evidence, the data set is rich enough to allow us to complement the study by Chuhan et al. (1996) in three respects. *First*, we draw attention to the interplay between the conditional volatilities of capital flows. Rather than analyzing the flows as such, we are interested in the dynamics and the interplay between their second moments, i.e., between the volatility of cross-border capital movements. *Second*, we do not only investigate the causality patterns between volatility series in a bilateral context but also apply a multivariate testing procedure.[26] *Third*, our data set covers a longer period of time, which

---

[26]   Chuhan et al. (1996) investigate the time series properties of capital flows in a multivariate framework by coding up a higher-dimensional vector autoregression. However, they do not explicitly address the question whether causality patterns between capital flows detected in a bivariate setting carry over to, e.g., a trivariate model.

begins in 1971 and ends in 1998. Chuhan et al. (1996) mention that the liberalization of global capital markets which took place in the 1980s might have altered
the dynamics of global capital flows. Upon resorting to a longer sample period, it
is possible to split up the data set and to examine whether or not the impact of the
liberalization of cross-border capital movements has indeed altered the interrela-
tion between volatilities of capital flows. Though Germany abolished most capital controls already in the late 1960s, the impact of the general trend towards a
liberalization of international capital flows might still have exerted an indirect
impact on the German capital account.

Quarterly data for the years 1971–1998 on net foreign direct investment, net
portfolio investment, and net short-term and net long-term credits (other investments) were collected from various issues of the balance of payments statistics
disseminated by the German Bundesbank. The seasonally adjusted German consumer price index was used to deflate the capital flow series. All capital flows are
expressed as a fraction of the seasonally adjusted German real gross domestic
product.

The method suggested by Schwert (1989) was employed in order to estimate
conditional volatilities of German capital flows. As compared to alternative esti-
mators, this method turned out to be fairly robust with respect to different model
specifications. The first step was to estimate a fourth-order autoregression for the
four categories of capital flows. The equations were enriched with four dummy
variables to allow for potentially differing quarterly intercepts. The same procedure was applied in a second step to the absolute values of the estimated residuals
of these equations. The fitted values of the equations estimated in the second step
provided an estimate of the conditional volatilities of the capital flows. A dummy
variable taking on the value minus one (plus one) in the third and fourth (first and
second) quarter of 1993 (1994) was incorporated into the model for portfolio and
other investments, and a dummy variable taking on the value minus one in the
fourth quarter of 1998 was added in the case of foreign direct investment. The
dummy variables were used to cushion the impact of single large outliers on the
estimated conditional capital flow volatility series. In the case of cross-border
lending, the above estimation approach was also applied to short-term and long-
term cross-border loans separately.

The resulting conditional volatilities are shown in Figure 3. In line with the
analysis of the coefficients of variation given in Table 5, other investments and
also short-term other investments exhibit several local peaks. In contrast, a similar clear-cut peak-trough pattern does not characterize the conditional volatility
of other long-term investments. This is a first hint that a differentiation between
short-term and long-term credits might indeed be important in empirical analyses
of the interplay between the volatility of various capital flows. Visual inspection
of the conditional variability of portfolio investments signals that the "base level"

Figure 3: Conditional Volatilities of German Net Capital Flows (1973:1–1998:4)

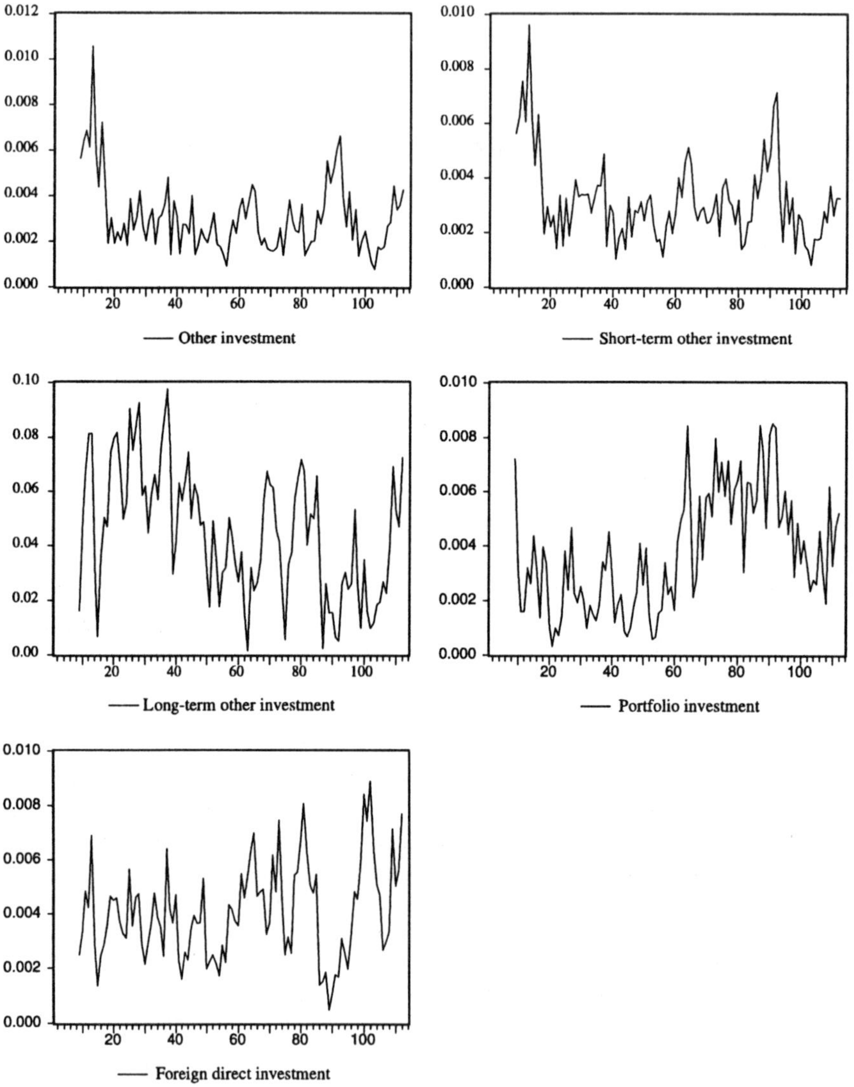

0.012
0.010
0.008
0.006
0.004
0.002
0.000
20
40
60
80
100
— Other investment
— Short-term other investment
0.10
0.08
0.06
0.04
0.02
0.00
— Long-term other investment
— Portfolio investment
— Foreign direct investment

of this volatility series rose slightly in the late 1980s and early 1990s. A similar proposition does not hold for FDI. Instead, Figure 4 shows that the amplitude of the peaks exhibited by this series, rather than the base level, increased modestly over time.

With the conditional volatility series at hand, we are now in a position to examine the interplay between the volatilities of various capital flows. An often-
used statistical technique to test for the predictive power of an economic variable
with respect to future changes in the level of another variable is the test for
Granger noncausality. Let the time series measuring the conditional volatility of
capital flow j in period t be denoted by σ_{tj}. Then the following bivariate autore-
gressive representation is estimated:

$$[7] \quad \begin{aligned} \sigma_{t,j} &= \alpha_0 + \sum_{i=1}^{s} \alpha_i \sigma_{t-i,j} + \sum_{i=1}^{s} \beta_i \sigma_{t-i,l} + \varepsilon_{t,j} \\ \sigma_{t,l} &= \gamma_0 + \sum_{i=1}^{s} \gamma_i \sigma_{t-i,l} + \sum_{i=1}^{s} \delta_i \sigma_{t-i,j} + \varepsilon_{t,l}, \end{aligned}$$

where the indices $j \neq l$ are used to denote FDI, PI, OI, short-term OI, and long-
term OI, and where $\varepsilon_{t,j}$ and $\varepsilon_{t,l}$ are normally distributed, mean-zero disturbance
terms. The lag length, s, is chosen using the minimum of the Schwarz information
criterion. The hypothesis that the conditional volatility of capital flow j does not
Granger-cause the conditional volatility of other cross-border capital movements
l (i.e., $\delta_i = 0$) can be tested by performing a standard F-test. The alternative hypothesis that the conditional volatility of capital flow l does not Granger-cause
the conditional volatility of capital flow j (i.e., $\beta_i = 0$) is tested in a similar fashion. If both hypotheses cannot be rejected, we have a feedback relationship.

Table 7 gives the results of the bivariate tests for Granger noncausality, which
reveal a fairly clear picture. The conditional volatility of FDI can be identified as
a major determinant of the variability of the other forms of capital flows. The null
hypothesis that FDI does not Granger-cause either other investment, portfolio investment, long-term or short-term other investments can clearly be rejected at
least at the 5 percent significance level. The data set also provides weak evidence
that the conditional volatility of other investment (short-term OI) tends to precede
the changes in the volatility of portfolio investment (PI). However, these latter
hypotheses can only be rejected at the 10 percent significance level and definitive
conclusions can be reached only after it has been assessed whether the results of
the bivariate tests for Granger noncausality are stable with respect to both the
specification of the sample period investigated and the dimension of the estimated models.

To address these questions, we first split up the sample period. As already
mentioned, Chuhan et al. (1996) pointed to the argument that the liberalization of
cross-border capital movements observed during the mid-1980s might have altered the dynamics of, and the linkages between, international capital flows.
Taking this line of argumentation into account, we reestimated the system for-

Table 7: Testing for Granger Noncausality between Volatilities of German Capital Flows (F-Tests for block exogeneity; bivariate model)

Time series		Lag length of the vector auto-regression	Schwarz information criterion	H_0: A does not Granger-cause B	H_0: B does not Granger-cause A
A	B				
1973:1–1998:4					
OI	PI	2	−20.51	2.38*	1.21
OI	FDI	1	−20.90	2.07	12.19***
PI	FDI	2	−19.21	0.31	5.06***
OI (short-term)	PI	3	−20.67	1.69	1.28
OI (short-term)	FDI	1	−21.01	2.82*	10.89***
OI (long-term)	PI	1	−15.23	2.44	2.24
OI (long-term)	FDI	1	−15.76	0.45	4.30**
1985:1–1998:4					
OI	PI	1	−20.85	0.21	0.03
OI	FDI	1	−21.18	1.07	16.19***
PI	FDI	1	−19.94	1.52	5.93*
OI (short-term)	PI	1	−20.85	0.92	0.52
OI (short-term)	FDI	1	−21.04	1.45	14.71**
OI (long-term)	PI	1	−15.14	0.13	1.09
OI (long-term)	FDI	1	−15.70	0.01	0.87

Note: ***(**,*) denotes that the null hypothesis of no causality is rejected at the 1 (5, 10) percent level. FDI = foreign direct investment; PI = portfolio investment; OI = other investment.

malized in [7] over the period 1985–1998. The results are documented in the lower part of Table 7. The results show that the leading position of FDI turns out to be largely robust with respect to the sample period chosen. Interestingly, and in contrast to the result obtained in the case of short-term credits, long-term other investments seem now to be largely unaffected by both portfolio capital flows and FDI. Although the volatility of FDI seems to be an important driving factor of the volatility of other forms of capital flows, a similar proposition cannot be made in the case of the volatility of long-term other investments. One possible explanation could be the fact that other investments include official capital flows, which are driven by other determinants than private capital flows. Moreover, the impact of the volatility of flows subsumed under the heading "other investment" ("short-term OI") on portfolio investment flows can no longer be detected.

As a final exercise, we checked whether the findings of the bivariate tests for Granger noncausality are confirmed when a third variable is added to the system of [7]. To accomplish this task, we estimated systems of simultaneous equations including the volatility of one of the OI series and the volatility of portfolio and foreign direct investment. The results of LR tests for block exogeneity are summarized in Table 8. Again, the causality patterns already elucidated by the results reported in the proceeding table show up: The volatility of foreign direct investment precedes changes in the volatility of credit flows. Given that this latter result is robust across several alternative testing procedures, we conclude that in the case of Germany and for the sample period under investigation the volatility of FDI has been an influential determinant of the fluctuations exhibited by other capital account items, in particular short-term capital.

Table 8: Testing for Granger Noncausality between Volatilities of German Capital Flows (LR Tests for block exogeneity; trivariate model)

Volatilities included in the vector autoregression	LR tests for the exclusion of lagged volatilities of		
	OI	PI	FDI
OI, PI, FDI	3.71 (0.15)	1.60 (0.45)	16.59 (< 0.01)
	OI (short-term)	PI	FDI
OI (short-term), PI, FDI	3.92 (0.10)	1.53 (0.46)	11.42 (< 0.01)
	OI (long-term)	PI	FDI
OI (long-term), PI, FDI	1.86 (0.39)	1.17 (0.56)	3.73 (0.15)
Note: Figures in parentheses denote marginal significance levels. According to the Schwarz information criterion, only one lag of the dependent variables was included in the vector autoregressive systems. FDI = foreign direct investment; PI = portfolio investment; OI = other investment.			

To provide an impression of the dynamic impact of FDI volatility on the conditional volatility of OI and PI, Figure 4 presents impulse-response functions capturing changes in the volatility of these latter series to a one-standard deviation shock to the volatility of FDI. The impulse-response functions are obtained

Figure 4: Selected Impulse-Response Functions for Volatilities of German Net Capital Flows

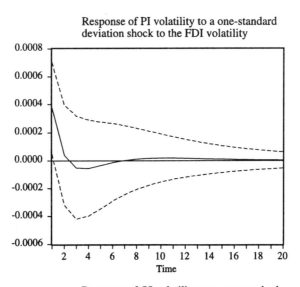

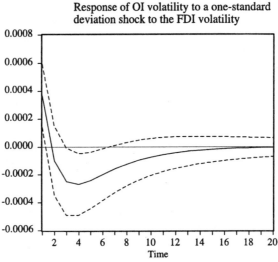

Note: The impulse-response functions are obtained from a trivariate model containing foreign direct investment (FDI), portfolio investment (PI), and other investment (OI). Dashed lines represent ±2 standard error bounds obtained from a Monte Carlo experiment with 1,000 simulation runs. The ordering of the variables used to obtain a unique Choleski decomposition is: FDI, PI, OI.

by estimating a trivariate VAR including FDI, PI, and OI, which was also the order of variables selected to obtain a unique Choleski decomposition. Figure 4 shows that the most pronounced effect of a shock to the volatility of FDI on the other conditional volatility series occurs in the first four quarters. The effect dies out after a few quarters, and the conditional volatilities of PI and OI revert to their respective pre-shock levels. This is in line with results obtained for the volatility of the returns of other financial market prices (see, e.g., Schwert 1989) and, thus, further corroborates the validity and robustness of our results.

By and large, these results confirm the finding of Chuhan et al. (1996) that the volatility of FDI tends to precede changes in other components of the capital account. While the study conducted by Chuhan et al. (1996) was devoted to an investigation of the interplay between cross-border capital flows per se, our empirical evidence provided above sheds light on the dynamic relationships between the volatilities of capital flows. Thus, the fact that we are able to confirm the results reported by Chuhan et al. (1996) with respect to the leading position of FDI relative to PI and OI is encouraging and indicates that more work in this area of research may yield further interesting and important results. Certainly, one important direction for future research could be to investigate whether or not the central role of the volatility of FDI flows can also be detected when the empirical methods employed in this paper are also applied to data for other countries.

c.   Capital Account Reversals

Finally, in order to assess the welfare implications of financial integration, not only the (short-run) volatility of capital flows matters but also the incidence of large current account reversals. If fluctuations of (net) capital flows cancel out quickly, implications for the real sector may be modest. If, however, capital flows swing for a prolonged period, policy adjustment may be necessary but may not be forthcoming as quickly as needed.[27]

Mussa et al. (1999) define a reversal of capital flows as a reduction in net capital inflows by more than 3 or 6 percent of GDP. Looking at the data for 17 emerging economies, which accounted for 75 percent of capital flows to emerg-

---

[27] It is beyond the scope of the present subsection to analyze whether capital flow reversals are triggered by the implementation of a single suboptimal policy measure at the onset of financial market liberalization or by a suboptimal policy reversal, which takes place at a later stage of the process of the deregulation of cross-border capital movements. See Eicher et al. (2000) for a theoretical discussion of this issue.

ing economies for the 1970s through 1990s, they find that large reversals became more frequent in the 1990s.

Figure 5 reproduces the results of Mussa et al. (1999) for our sample of OECD countries. We plot both reversals of net and gross flows, using a three- and a six-percent threshold level (relative to GDP). When using the three-percent threshold, reversals of net capital flows have appeared relatively frequently, and no clear trend for a clustering of large reversals is evident. As regards gross capital flows, relatively more large reversals were observed in two periods, in the early 1980s and the early 1990s. These are also the two periods which have seen very large reversals of capital flows, exceeding the six-percent threshold. Moreover, the number of countries affected by large capital reversals seems to have increased in the 1990s relative to the 1980s, at least when gross flows are considered. This largely supports the evidence documented for developing countries.

In addition, it should be noted that there is some evidence for increased large *inflows* of foreign capital into developed countries in the late 1990s. While emerging markets witnessed quite substantial outflows of foreign capital in the second half of the decade, the reverse holds true for OECD countries, which witnessed quite substantial increases in capital inflows during this time. In this sense, developments in the OECD area mirror developments in emerging markets, thus reflecting a redirection of international capital flows.

4.   Summary and Conclusions

The purpose of this paper has been to review the literature on global capital flows. In view of the variety and the broadness of the issues at hand, we have restricted the analysis to two main questions: First, what are meaningful concepts to measure capital mobility and the integration of international financial markets? Second, what changes in the structure and in the volatility of capital flows do we observe?

As regards the measurement of capital mobility and financial integration, it has been argued that none of the standard tests should be dismissed a priori but that different measures must be combined in order to obtain a comprehensive picture. Most tests do in fact point into the same direction of increased capital mobility over time. This paper has supported this general conclusion by looking at the time trend of gross capital flows over GDP. By this measure, the second half of the 1990s witnessed a particularly sharp increase in capital flows. Growth in GDP per capita, increased trade links, deregulation of capital flows both at the European level and internationally, and an increasing size of domestic financial

Figure 5: Number of OECD Countries Experiencing Large Capital In- and Out-flows, 1978–1998

a) Increase in capital outflows (three percent threshold)

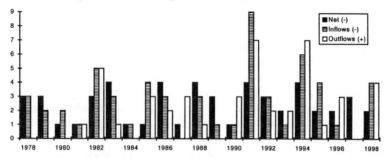

b) Increase in capital outflows (six percent threshold)

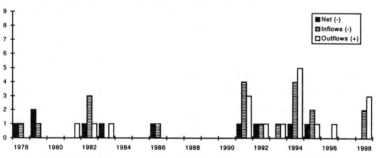

c) Increase in capital inflows (six percent threshold)

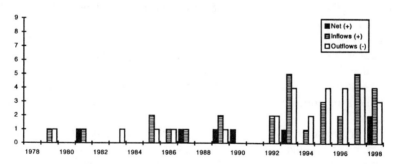

Note: Increased capital outflows have been defined in three different ways: as a reduction in net inflows [Net(–)], as reduction in gross inflows [Inflows(–)], and as an increase in gross outflows [Outflows(+)]. The same distinction can be made for increased capital inflows.

Source: IMF (2000), own calculations.

sectors are behind these trends. At the same time, the share of capital flows relative to GDP remained below 20 percent for most OECD countries.

In addition, the paper has shown that the structure of capital flows has changed over time. On a most general level, securitized finance tends to become more important as countries develop while the reliance on foreign direct investment tends to become less so. In this sense, the pattern of capital flows to developed and developing countries is consistent with the pecking order theory of international capital flows. At the same time, differences in the external financial structures of OECD countries remain substantial, and the causes of these differences have not yet been unearthed by the empirical literature. Hence, as regards routes for future research, the paper has shown a need for a better understanding of the factors determining the structure of international capital flows, in particular by taking account of the impact of asymmetries in information.

The results of the analysis of the interplay between various forms of German cross-border capital flows provides further support for this line of argumentation. Confirming the findings reported in earlier empirical investigations, the volatility of foreign direct investment was found to precede changes in the volatility of portfolio and short-term other investment. In contrast, the volatility of German long-term other investment turned out to be rather insensitive to variations in the volatility of either portfolio or foreign direct investment.

While we have presented evidence for German data only, both the robustness of our results and the fact that the evidence derived in the present paper is in line with the findings of earlier studies suggest that future examinations in this area of research using a more comprehensive data set could provide further interesting and important results. In particular, it would be interesting to explore whether the volatility of FDI can also be isolated as a leading variable with respect to the volatility of other capital account items when balance of payments data for other developed and emerging economies are utilized. More evidence on this subject would also be important from a theoretical point of view because the fact that FDI tends to precede other capital flows can be taken as support for models that assume that foreign direct investors acquire superior information by being present in the host country. If this holds true, their behavior can be taken as a signal by creditors or portfolio investors and thus trigger responses of other capital flows.

Work presented in this paper has no direct policy implications. Indirectly, however, it can provide some interesting insights. *First*, as regards the degree of integration into international capital markets, it has been argued that standard measures of integration provide relatively little information on the welfare effects of integration. There is, for instance, no direct link between the integration into international financial markets and the efficiency and contestability of the domestic banking system. Hence, policy makers aiming at greater financial openness of their economies may be well-advised to closely link internal and external financial liberalization.

Second, links between different types of capital flows are complex and there is, by implication, no optimal financial structure. Short-term bank lending and FDI may be more important at early stages of economic development and fulfill their specific economic functions, while securitized finance is likely to increase in importance later on. Yet, substantial cross-country variation in this pattern has also been evident. Without analyzing the links and the sequence of financial flows, it is difficult to argue that economic policy should aim at tilting the structure of capital flows into one or the other direction.

Similar conclusions hold, *finally*, for the volatility of financial flows. While FDI generally tends to be more stable than other forms of capital flows, changes in FDI also tend to precede increased volatility of other capital account items. Although research into the dynamics of capital flows has only just begun, this suggests that policies aimed at increasing or reducing the share of a particular capital flow must also take feedback effects between capital flows into consideration. In addition, before drawing policy conclusions, more evidence would be needed on the links between the volatility of capital flows, on the one hand, and real sector developments, on the other hand. This issue, however, has been beyond the scope of the present paper.

Appendix

Appendix A1: Data Definitions and Sources

Variable	Definition	Source
Panel data		
Control1	Dummy for the imposition of capital controls (= 0 for countries without restrictions, = 1 for countries with restrictions on capital account transactions)	IMF (1999)
Control2	Dummy for the imposition of capital controls (= 0 for countries without restrictions, = 1 for countries with different exchange rates for capital and current account transactions)	IMF (1999)
EU	Dummy variable for EU members (= 0 for nonmembers, = 1 for members without full compliance with EU capital account directive, = 2 for members with full compliance with EU capital account directive)	
Exports	Merchandise exports in billions of U.S. dollars (1998)	IMF (2000)
FDI	FDI in- and outflows, in millions of U.S. dollars, balance of payments basis	IMF (2000)
Financial centre	Dummy variable for financial centers (Bahamas, Great Britain, Hong Kong, Ireland, Luxembourg, Singapore, Switzerland) (= 0 for nonfinancial centers, = 1 for financial centers)	
Growth	Growth rate of real GDP volume index	IMF (2000)
real GDP per capita	Gross domestic product in billion current national currency, converted into U.S. dollars with the average annual exchange rate of the national currency to the U.S. dollar, deflated with the U.S. consumer price index, divided by total population	IMF (2000)
IM	Merchandise imports in billions of U.S. dollars (1998)	IMF (2000)
M2 / GDP	Broad money (billions of national currency, end-1998)	IMF (2000)

Appendix A1: Continued

Variable	Definition	Source
Other investment	Other investment in- and outflows, in millions of U.S. dollars, balance of payments basis	IMF (2000)
OECD	Dummy variable for OECD membership (= 0 for nonmembers, = 1 for members)	
Portfolio investment	Portfolio investment in- and outflows, in millions of U.S. dollars, balance of payments basis	IMF (2000)
Time series data		
Foreign direct investment	Foreign direct investment in- and outflows, in millions of U.S. dollars	Deutsche Bundesbank (2000)
Portfolio investment	Portfolio investment in- and outflows, in millions of U.S. dollars	Deutsche Bundesbank (2000)
Other investment	Other investment in- and outflows, in millions of U.S. dollars	Deutsche Bundesbank (2000)
Real GDP	German real gross domestic product; seasonally adjusted	IMF (2000)
CPI	German consumer price index; seasonally adjusted	IMF (2000)

Appendix A2: Country Sample

Australia	Iceland	Portugal
Austria	Ireland	Spain
Canada	Italy	Sweden
Denmark	Japan	Switzerland
Finland	Netherlands	United Kingdom
France	New Zealand	United States
Germany	Norway	

Bibliography

Bagnai, A., and S. Manzocchi (1996). Unit Root Tests of Capital Mobility in the Less Developed Countries. *Weltwirtschaftliches Archiv* 132(3):544–557.

Bartolini, L., and A. Drazen (1997). Capital Account Liberalization as a Signal. *American Economic Review* 87(1):138–154.

Bayoumi, T. (1990). Saving-Investment Correlations—Immobile Capital, Government Policy, or Endogenous Behavior? *IMF Staff Papers* 37(2):360–387.

—— (1999). Is There a World Capital Market? In H. Siebert (ed.), *Globalization and Labor*. Tübingen: Mohr Siebeck.

Bayoumi, T., and A. Rose (1993). Domestic Savings and Intra-national Capital Flows. *European Economic Review* 37 (6):1197–1202.

Bellak, C. (1996). International Capital Mobility—A Note. *Journal of International Money and Finance* 15(5):825–828.

BIS (Bank for International Settlements) (1999). *The BIS Consolidated International Banking Statistics.* Various Issues. Basle.

Bordo, M.D., B. Eichengreen, and J. Kim (1998). Was There Really an Earlier Period of International Financial Integration Comparable to Today? Working Paper 6738. National Bureau of Economic Research (NBER), Cambridge, Mass.

Breitung, J., and R. Brüggemann (1999). Uncovered Interest Parity—What Can We Learn from Panel Data? Mimeo.

Buch, C.M. (1999). Chilean-Type Capital Controls: A Building Block of the New International Financial Architecture? Kiel Discussion Paper 350. Kiel Institute of World Economics, Kiel.

—— (2000a). Why Do Banks Go Abroad? — Evidence from German Data. *Journal of Financial Markets, Institutions, and Instruments* (forthcoming).

—— (2000b). Financial Market Deregulation in the US—Lessons for Europe? Mimeo. Kiel Institute of World Economics, Kiel.

Buch, C.M., and S.M. Golder (2000). Disintermediation and Foreign Competition: No Threat to the German Banking System? *Banca Nazionale del Lavoro (BNL) Quarterly Review* (forthcoming).

Buch, C.M., and S. Lapp (1998). The Euro—No Big Bang for Financial Markets? *Konjunkturpolitik* 47:11–78.

Buch, C.M., and L. Lusinyan (2000). Determinants of Short-Term Foreign Debt. Working Paper 994. Kiel Institute of World Economics, Kiel.

Calvo, G.A., and C.M. Reinhart (1999). When Capital Inflows Come to a Sudden Stop: Consequences and Policy Options. Mimeo. University of Maryland.

Chuhan, P., G. Perez-Quiros, and H. Popper (1996). International Capital Flows: Do Short-Term Investment and Direct Investment Differ? Policy Research Working Paper 1669. World Bank, Washington, D.C.

Claessens, S., M.P. Dooley, and A. Warner (1993). Portfolio Capital Flows: Hot or Cool? In S. Claessens and S. Goopta (eds.), *Portfolio Investment in Developing Countries*. Washington, D.C.: The World Bank.

Coakley, J., and F. Kulasi (1997). Cointegration of Long Span Saving and Investment. *Economics Letters* 54(1):1–6.

Coakley, J., F. Kulasi, and R. Smith (1995). The Feldstein-Horioka Puzzle and Capital Mobility. Discussion Paper in Economics 6/95. Birkbeck College. University of London.

Deutsche Bundesbank (2000). *Zahlungsbilanzstatistik*, various issues. Frankfurt a.M.

ECB (European Central Bank) (1999). *Possible Effects of EMU on the EU Banking Systems in the Medium to Long Term*. Frankfurt a. M.: ECB.

Edwards, S. (2001). Capital Flows and Economic Performance: Are Emerging Economies Different? This volume.

Eicher, T.S., S.J. Turnovsky, and U. Walz (2000). Optimal Policy for Financial Market Liberalizations: Decentralization and Capital Flow Reversals. *German Economic Review* (1):19–42.

Engle, R.F., and C.W.J. Granger (1987). Co-Integration and Error Correction: Representation. Estimating and Testing. *Econometrica* 55(2):251–276.

Feldstein, M., and C. Horioka (1980). Domestic Savings and International Capital Flows. *Economic Journal* 90(June):314–329.

Frankel, J.A. (1989). Quantifying International Capital Mobility in the 1980s. Working Paper 2856. National Bureau of Economic Research, Cambridge, Mass.

Frankel, J.A., and S.L. Schmukler (1996). Country Fund Discounts, Asymmetric Information and the Mexican Crisis of 1994: Did Local Residents Turn Pessimistic Before International Investors? Working Paper 5714. National Bureau of Economic Research (NBER), Cambridge, Mass.

French, K.R., and J.M. Poterba (1990). Japanese and U.S. Cross-Border Common Stock Investments. *Journal of the Japanese and International Economies* 4:476–493.

—— (1991). Investor Diversification and International Equity Markets. *American Economic Review* 81(2):222–226.

Gehrig, T. (1993). An Information-Based Explanation of the Domestic Bias in International Equity Investment. *The Scandinavian Journal of Economics* 95(1):97–109.

Golub, S.S. (1990). International Capital Mobility: Net Versus Gross Stocks and Flows. *Journal of International Money and Finance* 9(4):424–439.

Gordon, R.H., and A.L. Bovenberg (1996). Why Is Capital So Immobile Internationally? Possible Explanations and Implications for Capital Income Taxation. *American Economic Review* 86(5):1057–1075.

Grilli, V., and G.M. Milesi-Ferretti (1995). Economic Effects and Structural Determinants of Capital Controls. IMF Working Paper 95/31. International Monetary Fund, Washington, D.C.

Gundlach, E., and S. Sinn (1992). Unit Root Tests of the Current Account Balance: Implications for International Capital Mobility. *Applied Economics* 24: 617–625.

Helliwell, J.F., and R. McKitrick (1998). Comparing Capital Mobility Across Provincial and National Borders. Working Paper 6624. National Bureau of Economic Research (NBER), Cambridge, Mass.

Hoffmann, M. (1999). The Feldstein-Horioka Puzzle and a New Measure of International Capital Mobility. Discussion Paper in Economics and Econometrics 9916. University of Southampton.

Hull, L., and L.L. Tesar (2000). The Structure of International Capital Flows. National Bureau of Economic Research (NBER). Working Paper. Victoria University of Wellington, New Zealand, and University of Michigan, Ann Arbor, USA.

—— (2001). The Structure of International Capital Flows. This volume.

Im, K.S., M.H. Pesaran, and Y. Shin (1997). Testing for Unit Roots in Heterogeneous Panels. Revised version of the DAE Working Paper 9526. University of Cambridge.

IMF (International Monetary Fund) (1998). *Balance of Payments Statistics Yearbook 1998. Part 2: World and Regional Tables*. Washington, D.C.

— (1999). *Exchange Restrictions*. Various Issues. Washington, D.C.

— (2000). *International Financial Statistics on CD-Rom (IFS)*. May. Washington, D.C.

Kellermann, K., and C.-H. Schlag (1999). Eine Untersuchung der Ersparnis-Investitions-Korrelation in Deutschland. *Zeitschrift für Wirtschafts- und Sozialwissenschaften* 119(1):99–121.

Kilka M. (1998). *Internationale Diversifikation von Aktienportfolios: Home Bias in Kurserwartungen und Präferenzen*. Frankfurt a.M.: Lang.

Kim, W., and S.-J. Wei (1999). Foreign Portfolio Investors Before and During A Crisis. Working Paper 6968. National Bureau of Economic Research, Cambridge, Mass.

Krol, R. (1996). International Capital Mobility: Evidence from Panel Data. *Journal of International Money and Finance* 15(3):467–474.

Laban, R.M., and F.B. Larrain (1997). Can a Liberalization of Capital Outflows Increase Net Capital Inflows? *Journal of International Money and Finance* 16(3):415–431.

Lane, P., and G.M. Milesi-Ferretti (1999). The External Wealth of Nations: Measures of Foreign Assets and Liabilities for Industrial and Developing Countries. IMF Working Paper 99/115. International Monetary Fund, Washington, D.C.

—— (2001). External Capital Structure: Theory and Evidence. This volume.

Lapp, S. (1999). Die Anlageentscheidungen deutscher Investoren — Der empirische Befund. Kiel Working Paper 937. Kiel Institute of World Economics, Kiel.

Lemmen, J. (1998). *Integrating Financial Markets in the European Union*. Cheltenham: Edward Elgar.

Levin, A., and C.-F. Lin (1993). Unit Root Tests in Panel Data: Asymptotic and Finite-Sample Properties. Unpublished Working Paper (revised). University of California San Diego.

Lewis, K.K. (1999). Trying to Explain the Home Bias in Equities and Consumption. *Journal of Economic Literature* 37:571–608.

Loungani, P., A. Razin, and C.-W. Yuen (2000). Capital Mobility and the Output-Inflation Tradeoff. IMF Working Paper 00/87. International Monetary Fund, Washington, D.C.

Mishkin, F. (1998). International Capital Movements, Financial Volatility, and Financial Instability. Working Paper 6390. National Bureau of Economic Research, Cambridge, Mass.

Montiel, P.J. (1994). Capital Mobility in Developing Countries: Some Measurement Issues and Empirical Estimates. *The World Bank Economic Review* 8/3:311–350.

Mussa, M., A. Swoboda, J. Zettelmeyer, and O. Jeanne (1999). Moderating Fluctuations in Capital Flows to Emerging Market Economies. Paper presented on the Conference on Key Issues in Reform of the International Monetary and Financial System, May 28–29, 1999. mimeo.

Myers, S.C., and N. Majluf (1984). Corporate Financing and Investment Decisions When Firms Have Information Investors Do Not Have. *Journal of Financial Economics* 13:187–221.

Niehans, J. (1994). Elusive Capital Flows: Recent Literature in Perspective. *Journal of International and Comparative Economics* 3:21–43.

Obstfeld, M. (1995). International Capital Mobility in the 1990s. In P.B. Kenen (ed.), *Understanding Interdependence — The Macroeconomics of the Open Economy*. Princeton, New Jersey: Princeton University Press.

Obstfeld, M., and K. Rogoff (1996). *Foundations of International Macroeconomics*. Cambridge, Mass.: MIT Press.

Obstfeld, M., and A.M. Taylor (1997). The Great Depression as a Watershed: International Capital Mobility Over the Long Run. Working Paper 5960. National Bureau of Economic Research, Cambridge, Mass.

Portes, R., and H. Rey (1999). The Determinants of Cross-Border Equity Flows. Working Paper 7336. National Bureau of Economic Research, Cambridge, Mass.

Razin, A., E. Sadka, and C.-W. Yuen (1998). A Pecking Order of Capital Flows and International Tax Principles. *Journal of International Economics* 44:45–68.

Rodrik, D., and A. Velasco (1999). Short-Term Capital Flows. Working Paper 7364. National Bureau of Economic Research, Cambridge, Mass.

Sarno, L., and M.P. Taylor (1999). Hot Money, Accounting Lables and the Permanence of Capital Flows to Developing Countries: An Empirical Investigation. *Journal of Development Economics* 59:337–364.

Schwert, G.W. (1989). Why Does Stock Market Volatility Change Over Time? *The Journal of Finance* 44:1115–1153.

Shibata, A., and M. Shintani (1998). Capital Mobility in the World Economy: An Alternative Test. *Journal of International Money and Finance* 17:741–756.

Sinn, S. (1992). Saving-Investment Correlations and Capital Mobility: On the Evidence from Annual Data. *Economic Journal* 102 (September): 1162–1170.

Taylor, A.M. (1996). International Capital Mobility in History: The Savings-Investment Relationship. Working Paper 5743. National Bureau of Economic Research, Cambridge, Mass.

Tesar, L., and I.M. Werner (1992). Home Bias and Globalization of Securities. NBER Working Paper 4218. National Bureau of Economic Research, Cambridge, Mass.

von Furstenberg, G.M. (1998). From Worldwide Capital Mobility to International Financial Integration: A Review Essay. *Open Economies Review* 9:53–84.

William R. Cline

The Management of Financial Crises

1.   Introduction

The late 1990s was marked by a series of financial crises in emerging market economies, notably including Mexico (1995), East Asia (Indonesia, Korea, Malaysia, Philippines, and Thailand in 1997–1998), Russia (1998), and Brazil (1998–1999), as well as a number of smaller economies by 1999 (Pakistan, Ecuador, Ukraine, Romania). The manner of dealing with these crises has engendered heated debate about the proper macroeconomic policies for countries entering crisis to adopt, the magnitude and conditions of international official sector support, and the appropriate nature of involvement of the private sector in resolving crises.

After an initial synoptic review of the issues regarding country policy responses, this study examines the analytics of, and evidence on, the relative merits of alternative approaches of both the official international sector and the private sector in resolving crises. The discussion gives special attention to the issue of involvement of the private sector in crisis resolution. The central theme is that the nature of involvement will vary case by case, but the organizing principle should be to mobilize private sector participation in as voluntary a way as possible under the circumstances. This will maximize the chances for speedy return to private capital market access. This objective is especially important for prospects for economic development in a world economy in which private rather than official capital flows have become overwhelmingly the dominant source of external savings for economic growth.

The financial crises have typically involved the following features: an intensifying runoff in short-term external debt and, correspondingly, a rapid loss in reserves; a severe drop in the exchange rate (on the order of 40 percent); and a

Remark: The views expressed here are those of the author and should not be interpreted as official positions of the Institute of International Finance.

collapse in economic growth for a year or more following the crisis. Whether external debt restructuring has been required has depended primarily on whether the crisis was essentially one of short-term illiquidity (Mexico, Korea, Thailand, Brazil) or longer-term quasi-insolvency problems driven by relatively higher total external debt and/or more fundamental political breakdowns (Russia and Indonesia; Ecuador, Pakistan, Ukraine).

This paper does not address the important range of issues associated with crisis prevention. However, it should be emphasized that progress has been made in this area, particularly in data transparency (see IIF 1999b), IMF disclosure of country surveillance results (publication of some Article IV reviews), and a significant shift toward floating exchange rates.

2.   Country Adjustment Policies

A forceful economic adjustment program is an essential ingredient of crisis resolution, as it is the key to restoring confidence in private capital markets. Exchange rate and monetary policy adjustment were essential in Mexico. Fiscal adjustment was central in Brazil, and its absence was the driving force in the breakdown of orderly adjustment in Russia. In East Asia, where fiscal accounts had been stronger, the primary challenges were in the structural areas of domestic banking system fragility and (in Korea) correcting excessive corporate leveraging and opening domestic capital markets to foreign direct and portfolio equity. Despite the vociferous critiques of IMF-supported adjustment strategies in the region (notably by Stiglitz 2000), their temporarily tight monetary policies and high interest rates were necessary to stem the collapses of exchange rates.[1] The dramatic recovery in Korea in particular by 1999–2000 surely should count as supportive evidence, considering that if recession had persisted instead it would undoubtedly have been cited by the critics as evidence to the contrary.

---

[1]   There is a better case that the initial targets for fiscal targeting were in error, and the IMF has acknowledged as much. Neither the governments nor the IMF anticipated the severity of the recessions, and hence the risks of procyclical fiscal tightening. For an early diagnosis along these lines, along with an analysis that otherwise finds the adjustment programs (including their temporary monetary tightening) broadly appropriate, see Cline (1998).

3.   Contemporary Capital Markets

To develop a meaningful analysis of private-sector involvement in crisis resolution, it is first necessary to understand the nature of the international capital markets facing emerging market economies today. At the outset of the 1980s before that decade's debt crisis, capital flows were mainly in the form of syndicated bank loans to sovereigns. After the debt crisis, banks tended to limit lending to Latin America to short-term trade credit, and bonds became the principal vehicle for capital flows. Bank lending did grow vigorously in Asia prior to the 1997–1998 crisis. Much of it went to local banks and corporations, rather than to sovereigns (which, unlike their Latin counterparts, were not much in deficit and so did not need to borrow). As credit flows were shifting from banks as the source and sovereigns as the borrowers toward bonds as the source and much higher participation by private borrowers, equity flows through portfolio and especially direct investment were increasing persistently until the latter came to dominate overall capital flows.

As Figure 1 shows, the crises of the late 1990s sharply reduced net bank lending to emerging markets.[2] Net bond flows (in "Other private creditors") and equity held up better, and direct investment continued to rise despite the crises. As Figure 2 shows, by 1998 bonds and other nonbank credit accounted for more than half of debt to private creditors, whereas a decade earlier banks had accounted for more than two-thirds.[3] The figure also shows the diminishing share of short-term bank claims from their peak in 1996, when they had built up to excessive levels in East Asia in particular (nearly $100 billion in Korea at a time when its external reserves widely defined were only $34 billion).

Policy toward financial crisis resolution must face foremost the reality that in the contemporary capital market, it is market expectations that will dominate capital flows, rather than official mandates. This is a contrast to 1985, when U.S. Treasury Secretary James Baker could outline the Baker Plan whereby the banks would commit to several billion dollars of new lending to Latin America. There has been a shift from "customer market" financial flows—concentrated in a few large banks with ongoing relationships with sovereigns—to "financial commodity market" flows through bonds, portfolio equity, and direct equity investments.

---

[2]   Emerging market totals in Figure 1 and elsewhere in this paper refer to the 29 largest emerging market economies unless otherwise noted. See IIF (2000a).

[3]   See IIF (2000b) for aggregates in Figure 2, which refer to 37 emerging market economies.

Figure 1: Net Capital Flows to Emerging Markets

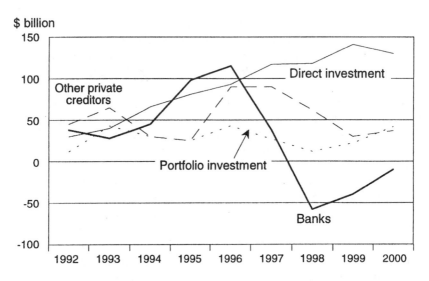

Figure 2: Emerging Market Debt to Private Creditors

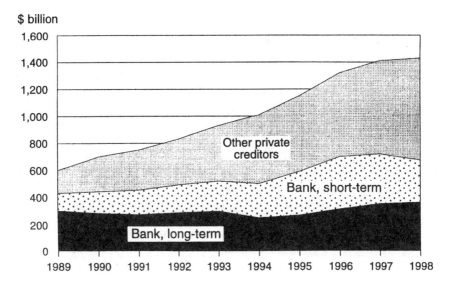

Many interpret these changes as evidence of a need for new debt rescheduling mechanisms for bonds, to complement the "London Club" bank reschedulings led by an advisory group of major banks involved. However, the more appropriate interpretation is that the stakes have substantially increased for resolving crises in a market-friendly way that restores confidence rather than locks in existing creditors. The discussion below returns to the issue of rescheduling bonds.

4.   Debt Theory

It is also necessary to return to basic debt theory if a policy toward crisis resolution is to have firm underpinnings. Probably the best construct for understanding sovereign debt remains that provided by Eaton and Gersovitz (1981). They asked why anyone would lend to a foreign government largely immune from the sanction of seizure of collateral that is present in most lending. Their answer was: sovereign borrowers seek to preserve reputation in order to be able to borrow again in the future. They modeled this in terms of the welfare value of consumption-smoothing through borrowing in a bad (e.g., low commodity price) year and repaying in a good one. The principle is broader, however.

The essence of sovereign lending, then, is that creditors must believe that the borrower will feel compelled to make every effort to service the debt because otherwise its loss of reputation will shut it out of the capital markets for some time to come. It follows that any proposals for international architecture must carefully consider the resulting impact on this "bonding" relationship. Arrangements that tend toward international political facilitation of default will tend to undermine the reputational sanction by giving sovereign debtors the impression that they can escape payment with little repercussion. Correspondingly, such arrangements will tend to raise doubts in potential creditors' minds about the value of the reputational sanction. The consequence will tend to be fewer loans, priced at higher interest rates to cover the increased risk. While some would say that less lending is precisely what is desirable after the excesses of the mid-1990s, they fail to be aware of the persistent dearth of lending, and persistent high interest rates, that have characterized emerging markets subsequent to the East Asian and Russian crises. Thus, as shown in Figure 1, projected net bank lending for 2000 remains negative in the aggregate; and as shown in Figure 3, in general the spreads in emerging market lending (interest rate premium above U.S. treasury obligations) remain considerably above their levels prior to the East Asian crisis.[4]

---

4   EMBI in Figure 3 refers to IP Morgan Emerging Markets Bond Index.

Figure 3: Bond Spreads

Basis points

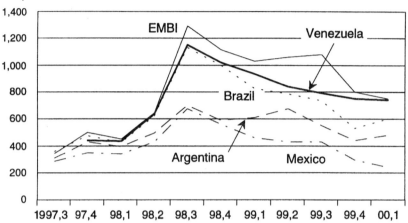

A second analytical construct would also seem useful in considering the appropriate means for involving the private sector in crisis resolution. This is the concept of "market based collective action" (IIF 1999a: 11). In terms of game theory, large international banks may be seen as players in a "repeat game" that is positive-sum. By cooperating in maintaining their short-term credit lines (as they did in Brazil), each major player is establishing its behavior pattern for each of the others to judge likely behavior in a similar future situation. Bondholders are too numerous, small, and dispersed to be seen as repeat-game players, and hence are unlikely to be parties to such arrangements. There will often nonetheless be enough of a critical mass of short-term bank claims that the undertaking of such an arrangement boosts confidence as well among bondholders (whose claims, in any event, are payable over some years rather than on demand).

5.   Success Stories: Mexico, Korea, and Brazil

With these considerations in mind, it is useful to turn to the actual experience in the key crisis cases of the last few years. Judged by the objectives of preserving systemic stability, accomplishing speedy return to access to private capital markets, and achieving a return to economic growth, the three most important cases also stand out as successes: Mexico in 1995, Korea in 1997–1998, and Brazil in

1998–1999. Moreover, although in each case there were large amounts of public sector funds committed in principle, in all three these funds were repaid or being repaid within a relatively short time.

Figure 4 shows net private capital flows to Mexico, Korea, and Brazil before, during, and after the year of each of their respective crises.[5] The success in prompt return to capital markets is shown by the rebound from negative to positive flows by the first year after the crisis (t+1), in the cases of Mexico and Brazil, and by the second in the case of Korea. Moreover, in Korea already by mid-1998, following the outbreak of the crisis in late 1997, negative private flows represented reduced demand for foreign credit at least as much as reduced supply, as domestic corporations cut back excessive leveraging.

Figure 4: Private Capital Flows before and after Market-Based Crises Resolution

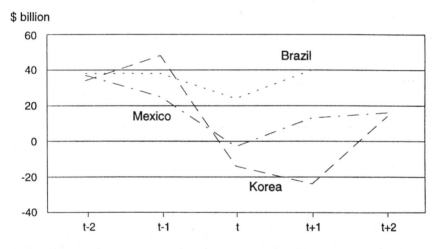

Growth recovery was also favorable in each of these cases. In Mexico real GDP growth fell from 4.4 percent in 1994 to –6.2 percent in 1995 but rebounded to 5.2 percent in 1996 and 6.8 percent in 1997. In Korea, growth fell from 5 percent in 1997 to –5.8 percent in 1998 but surged to 10 percent in 1999 and was projected at 8 percent in 2000. In Brazil, growth had already fallen from

---

5   The crisis year t is 1995 for Mexico, 1997 for Korea, and 1999 for Brazil. Data are from Institute of International Finance estimates.

3.6 percent in 1997 to –0.1 percent in 1998, recovered mildly to 0.8 percent in 1999, and was expected to reach 4.5 percent in 2000.[6]

Similarly, in all three cases there was a turnaround from temporary use of official international borrowing to repayment. Thus, Figure 5 essentially shows the inverse of the pattern for private capital flows. Public inflows were minimal before the crisis year, turned relatively large in the year of the crisis (the highest being net inflow of $25 billion to Mexico in 1995), and then reversed to large repayment outflows after the crisis (on the order of $10 billion annual net repayments in each of the three cases).

Figure 5: Net Public Capital Flows in Three Crises

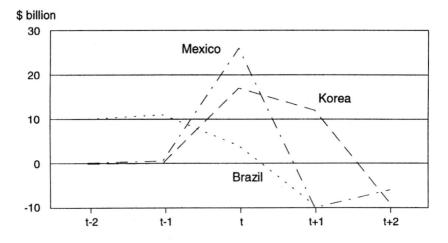

This inverse relationship suggests that public support was playing a crucial "balance wheel" role in contributing to adjustment programs that restored confidence in the first year of the crisis and setting the stage for repayment soon thereafter. This type of role has become familiar in other areas of public economic policy (natural disaster relief, countercyclical macroeconomic policies). Yet concerns about moral hazard (discussed below) have clouded what should be a crystal clear historical evaluation: international economic policy succeeded dramati-

---

6   The actual rates for 2000 turned out to be 9 percent in Korea and 4.2 percent in Brazil.

cally in stemming systemic risk and reestablishing the conditions for renewed market access and growth in the crisis cases of Mexico, Korea, and Brazil. Much of the debate on crisis resolution would be better directed if it focused on what went right in these cases rather than on how in these and other crises the official sector should somehow have imposed greater burdens on private creditors.

Nor should there be much doubt that these were systemic cases. The combined external debt of these three countries alone amounted to $536 billion at the end of 1996 (IIF 2000b), or more than one-fourth of the emerging market total. While industrial country bank exposure to emerging market economies was far lower by the mid-1990s than at the outset of the 1980s debt crisis, so that the potential risk to industrial country financial systems was not as direct as in the earlier episode, the stakes were nonetheless once again systemic. This is perhaps most evident in the severe swings in financial markets (witness the U.S. stock market collapse after the Russian default in August 1998) and trade flows (about $120 billion shift into trade surplus for East Asia alone) that did occur from emerging markets crisis contagion despite the substantial success in keeping it from becoming much worse.

In each of the three large success cases, there were strong positives that made it possible for the combination of international support, policy adjustment, and (in Korea and Brazil) private sector participation to overcome the crisis relatively quickly. Overall indebtedness was intermediate or, in the case of Korea, low, and the proximate problem was a liquidity squeeze caused by a mismatch between short-term external debt and reserves.[7] All three countries had experienced years of policy and structural improvements. As large, semi-industrial economies (and with strong direct investment inflows in the cases of Mexico and Brazil), all three were well positioned to overcome liquidity crises once policies were realigned (in Korea and Brazil, with the help of an improvement in domestic political coherence; and in Mexico and Brazil, once the paradigm of the quasi-fixed exchange rate was recognized to have failed).

---

[7] Even in Brazil and against a moderate export base of a large, relatively closed economy, external debt net of reserves stood at about 265 percent of exports of goods and services at end-1997, compared with almost 400 percent in 1982. Korea's end-1996 ratio of total external debt net of reserves to exports of goods and services was only 83 percent.

6.   Failures: Russia, Ecuador

In contrast, Russia and Ecuador constitute unambiguous failures of financial crisis resolution in the late 1990s.[8] In both cases, the key ingredient of forceful adjustment in domestic macroeconomic and structural policies was missing, fundamentally because of domestic political divisions. In both cases, fiscal deficits and
capital flight were persistent and large. To be sure, both countries also faced bad
luck in the collapse of oil prices in 1998 (and, in Ecuador, severe El Niño damage).

Relative indebtedness was not overwhelming in either case (external debt net
of reserves stood at about 200 percent of exports of goods and services at end-
1997 in Russia and 300 percent at end-1998 in Ecuador), but this was of little
help given the lack of coherence in domestic policy and capital flight. The decision to impose unilateral default was the most unambiguous mark of the failure of
crisis resolution in both countries.

There were other complicating factors. In the case of Russia, geopolitical
moral hazard had contributed to large foreign private capital inflows, even as
residents placed capital abroad. It became increasingly clear by mid-1998 that
policies such as extremely high treasury bill interest rates to roll over debt were
unsustainable. Market participants increasingly concluded that even the sizable
IMF support (under an adjustment program considered weak by many in the
market) would not suffice to offset escalating capital flight. As it further became
clear that the official international community was unprepared to redouble its
bets in the face of tepid adjustment progress, the stage was set for collapse. In
Ecuador's case, there was the complicating factor that official international entities apparently encouraged policy makers in the notion that rescheduling private
debt (including Brady and Euro bonds) would be required, in part because of the
policy of Paris Club "comparability" that if bilateral debt were rescheduled, private claims should be as well. The discussion below returns to this issue.

---

[8]   To sharpen distinctions, the discussion omits other arguable successes (Thailand,
Philippines, Malaysia) as well as other prime candidates for classification as cases of
failure (Indonesia).

7.   Voluntary versus Involuntary Private Sector Participation

The favorable outcomes for Mexico, Korea, and Brazil lend support to the principle suggested at the outset: that involvement of private creditors in crisis resolution should be on as voluntary a basis as possible, to assure the soonest possible return to normal capital market access. Table 1 lists a spectrum of approaches that have occurred in the main recent crisis cases, arrayed from most to least voluntary. The international support program for Mexico in 1995 did not involve a specific mechanism for private-sector involvement, in part because the proximate problem lay with short-term government paper (Tesobonos) rather than bank loans and the traditional response of appealing to leading banks for emergency maintenance of lending was not an option. The Mexican case was thus the most voluntary, and corresponded the most to a pure public-sector lender-of-last-resort intervention.

Table 1: Modalities of Private-Sector Involvement in Crisis Resolution

	Cases:
I. More voluntary	
Spontaneous private capital return	Mexico
Contingent credit lines	Argentina, Mexico
Maintenance of short-term credit lines	Brazil
Conversion of short-term claims to medium-term	Korea
Limited sectoral rescheduling	Thailand
II. Less voluntary	
Broad-negotiated rescheduling	Indonesia private debt, Latins in 1980s
Unilateral debt exchange offer before default	Pakistan, Ukraine
Unilateral default prior to exchange negotiation	Russia, Ecuador

Also strictly voluntary is the next item in the table, contingent credit lines prearranged on market terms and callable by the government in question in the event of an emergency. As a major example, Argentina maintains about $7 billion in such lines with about a dozen international banks. Commitment fees for such arrangements can be relatively expensive if they are undertaken in a period of weak confidence in emerging markets lending. Nonetheless, in a probabilistic sense they can be a good bargain, because their welfare benefit to the country is potentially very high (equal to their contributed reduction in the probability of a financial crisis multiplied by the GDP loss likely to occur in a crisis).[9]

The next entry in the table crosses the threshold from spontaneous to quasi-voluntary private sector collective action. It refers to the arrangement adopted by leading international banks in the second quarter of 1999 to maintain their credit lines to Brazil. This arrangement played a pivotal role in stemming capital outflows and severe pressure on the real.

The arrangement for Brazil demonstrated that even where bond and other non-bank claims are relatively large, a critical-mass nucleus of support from banks in maintaining short-term lines can contribute crucially to restoring confidence. With IMF assistance, the Brazilian government had already developed a system for monitoring trade and interbank short-term loans in early 1999. The government was thus in a position to reassure the key lenders about the status of adherence to the maintenance of lines (which began quite high, and was still relatively high by August when the government declared the arrangement was no longer needed). This information helped overcome the potential prisoner's dilemma problem (each player taking measures in isolation that would be counterproductive for itself and the group as a whole for lack of ability to communicate and form a superior joint strategy).

The Korean conversion agreement of early 1998 was somewhat more "quasi" and somewhat less voluntary. Even so, it was voluntary in the sense that there was no formal government blockage of payment to nonparticipants, and indeed in both cases a number of smaller lenders ran down credit lines rather than maintain (Brazil) or convert them (Korea). Over $20 billion in short-term credit to Korean banks was converted to 1- to 3-year paper backed by the government, at spreads in the range of 225–275 basis points, below market at the time of the crisis but considerably more remunerative than the original loan terms. Pressure from G-7 central banks at year-end 1997 played an important role in mobilizing participation by the largest banks, but the overall effect was a much more voluntary arrangement than a more sweeping and formal rescheduling of external debt. The relatively high concentration of debt in the form of short-term obligations to banks facilitated this resolution. This difference from Mexico's Tesobono prob-

---

9   For a discussion of private sector contingent credit lines, including the acrimony they caused in Mexico in late 1998 when the government drew on about $3 billion at terms the lenders considered by then to have become outdated by market events, see IIF (1999a: 35–37).

lem is a vivid illustration of why in implementation the best approach will likely vary from case to case.

Next in the table is the surgical restructuring of finance company debt in Thailand, where otherwise crisis resolution with foreign official support did not reschedule private claims or seek other special private sector involvement. The table then shifts to the increasingly involuntary solutions of broad debt rescheduling and, least voluntary, unilateral default. Indonesia's framework for private debt restucturings has not generated much in the way of agreements so far, in considerable degree because the weakness of domestic bankruptcy laws and procedures makes it difficult to bring corporate debtors to agreement. The recourse of Pakistan and Ukraine to unilateral offers to exchange Eurobonds, in part at the Paris Club's urging on grounds of comparability, and the even more unilateral approaches of preemptive default by Russia and Ecuador, are discussed below in the section on workouts. The Pakistan and Ukrainian deals lengthened rather than reduced the debt obligation and were thus less extreme than the Russian and Ecuadorian outcomes.[10] Even so, the further along this involuntary range in the spectrum of private sector participation, the more the expectation has to be that it will be a long time before normal capital market access is restored.

8.   Magnitude of Official Support

The principal financial crises of the late 1990s were marked by a new public sector phenomenon that reflected the new characteristics of capital markets: official support linked to policy adjustment programs was much larger, albeit on a short-term basis, than under typical IMF programs of the past. The headline figures of IMF, other multilateral support, and "second line of defense" bilateral support amounted to $50 billion for Mexico, $17 billion for Thailand, $34 billion for Indonesia, $16 billion for Russia, and $42 billion for Brazil (IIF 1999a: 48). For the IMF in particular, this meant much larger support than under traditional quota-linked formulas. Thus, the IMF support for Korea was set at $21 billion, or about 19 times quota—compared with more traditional ceilings on the order of three times quota.

---

[10]   A unilateral bond exchange offer to holders of Ecuadorian Brady and Eurobonds, involving significant forgiveness, was successfully carried out in August 2000 subsequent to the presentation of this paper.

The impression of massive public sector bailouts provoked political calls for "burden sharing" by private creditors. Ironically, throughout the 1990s, the vast bulk of financing of emerging markets was by the private sector, not the public sector. Of the $1.9 trillion in net capital flows that went to emerging markets during 1991–1999, $1.66 trillion, or 86 percent, was from the private sector (Figure 6). Even in 1997–1998, the peak period of official crisis-related support, net public flows to all emerging market economies amounted to only 19 percent of the total. In part this paradox is explained by the fact that actual disbursements of the large packages were considerably less than face value (only about half or somewhat less for Thailand and Korea by mid-1998, and only one-sixth for Indonesia by then [IIF 1999a: 54]).

Figure 6: Private versus Public Capital Flows to Emerging Markets

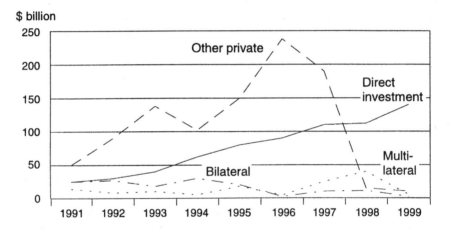

A study group for the Council on Foreign Relations has criticized the large lending programs for creating moral hazard and "provid[ing] cover for holders of short-term debt to escape the consequences of their poor lending decisions" (Council on Foreign Relations 1999: 16). It recommends that "in the vast majority of cases" IMF support should be limited to 100 percent of IMF quota annually and 300 percent cumulatively, although it does recognize that in cases of systemic threats larger lending may be appropriate.

I will argue below that the moral hazard concern has been exaggerated, and that the empirical evidence does not support the view that public sector bailouts have caused subsequently excessive private sector lending to emerging markets. If we set aside the moral hazard question then, for the moment, the case for much larger crisis-related lending than in the past seems compelling, and not only for systemic countries. The reason is that at an earlier time when capital controls were prevalent, and when the traditional ratios of IMF support to quota were determined, the current account deficit was the principal financing concern. In today's mobile capital markets, a liquidity crisis can involve a financing challenge that far exceeds the current account deficit. Indeed, in Korea's case, the prospective current account deficit in 1997 was only $8 billion (and there was a *surplus* of $40 billion in 1998).

The central issues are: how much of a show of official support does it take to restore market confidence; and, is there a real cost to the higher magnitudes of support needed? The G-7 answered these questions in practice not only by undertaking the large support programs just mentioned, but also by introducing a new facility in the IMF, the Supplemental Reserve Facility, that provides for far larger support relative to quota than traditionally, at penalty interest rates (300–500 basis points above normal IMF lending rates), to be repaid in a shorter period (12–18 months).

About the only theoretical construct for determining the proper amount of official lending in a crisis still remains Bagehot's (1873) famous proposition that in a panic, the central bank should lend without limit to solvent banks, but nothing at all to the insolvent. The late 1990s saw a broadly successful public sector application of Bagehot's rule to emerging markets, in an effort that sought not only to preserve the fledgling emerging capital markets after the seizing up of these markets in the 1980s debt crisis, but also to avoid wider potential threats to the global economy from a meltdown in these economies.

Aside from the potential moral hazard cost of such efforts, the principal potential cost to the public sector is the risk that Bagehot's test is not met: the country turns out to be insolvent, and there results a sequence of Paris Club reschedulings and eventually forgiveness, at a cost to the public sector. Figure 7 thus proposes the notion of a "Bagehot curve." On the horizontal axis, it shows the probability that the country is a case of insolvency. Unity represents an insolvent case likely to be unable to repay fully even debt owed to public sector lenders. Zero represents solely liquidity difficulties, with no question of their transiting into insolvency.

On the vertical axis, the prospective amount of public emergency financing that should be considered is shown. Intercept F_0 shows the traditional, limited IMF quota-linked amounts; intercept F_1 shows the much larger amounts in the Supplementary Reserve Facility and bilaterally linked packages. The intersections with the "Bagehot curve" show that when the probability of insolvency begins to be substantial (at for example P_B) and the emergency lending might face a serious risk of nonrepayment, official support is best left to the traditional modest

Figure 7: The Bagehot Curve

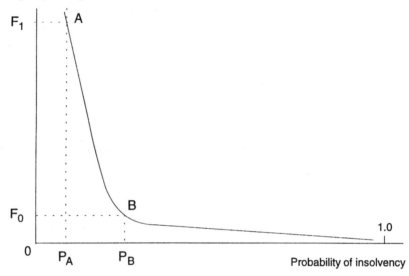

level F_0. If instead the probability of insolvency is very low at P_A, the Bagehot test is met and it is socially efficient to lend large public funds temporarily to restore private-sector confidence. Correspondingly, at intersection B, there will be much greater likelihood of rescheduling of private claims, with the "voluntary" element increasingly vanishing as the insolvency probability rises. At intersection A, there should be little if any need for private rescheduling, and some resolution at the upper end of the voluntary spectrum of Table 1 should suffice.

Finally, those (including Council on Foreign Relations 1999) who recommend that public sector financing that is large relative to IMF quota should only be incurred when the international financial system is at risk would seem to encourage "too big to fail" moral hazard for authorities of only the largest emerging market economies. It would seem much more equitable as well as efficient to make the availability of proportionately large emergency financing primarily a function of the country's location along the Bagehot curve (and the forcefulness of its adjustment measures), rather than of its economic size and potential for imposing a systemic threat.

9.   Moral Hazard

Of course, if there is a major distortion of private-sector incentives toward excessive lending to emerging markets as a result of the large emergency packages of the late 1990s, then the analysis should shift much more in the direction of limiting such financing and pressing the private sector along toward the involuntary end of the spectrum of participation if need be as a consequence. Considerable evidence suggests, however, that the concern about moral hazard has been exaggerated, and that to the extent this distortion of incentives is present, it is more than offset by other forces that have combined to curtail lending to emerging markets.

Estimates in IIF (1999a: 56–61) suggest that from mid-1997 through end-1998, private sector losses in East Asia and Russia amounted to some $60 billion for banks, $50 billion for bondholders, and $240 billion for foreign investors in equity markets. These large losses seem likely to have left the private sector with the impression that investing in emerging markets is risky indeed, instead of the impression that it is risk-free because of official bailouts.

Similarly, Zhang (1999) provides formal econometric tests of the moral hazard impact of the Mexican financial rescue on emerging markets. In equations explaining lending spreads (above U.S. treasury rates) for several emerging market economies by variables for global market conditions (represented by the spread on U.S. high-yield corporate bonds) and country economic performance (debt/export ratio, imports/reserves, and inflation), he finds that a dummy variable for the post-Mexican-crisis period (4th quarter 1995 through 2nd quarter 1997) has the wrong sign for the moral hazard hypothesis, and is statistically insignificant. The driving force behind falling spreads in this period was the buoyancy in the relevant high-risk global capital markets more generally, not Mexico moral hazard.[11]

The actual volumes of lending flows remain fragile and levels of spreads remain high in emerging markets, more than two years after the large support program for Korea and five years after that for Mexico. Net bank flows will likely be negative again this year, and lending spreads remain well above their pre-Asian-crisis levels for most emerging market economies (Figures 1 and 3). If the dominant force in determining lender expectations were moral hazard from official

---

[11]   In contrast, the Meltzer Commission (2000) and some others seem to take it as an article of faith that the surge in lending to emerging markets in 1996 and early 1997 resulted from Mexico moral hazard. However, there seem to be no formal tests supporting this view, which instead appears to rest on post hoc ergo propter hoc reasoning.

emergency support programs, surely prospective lending volumes would have rebounded much more and spreads narrowed further.

The evidence thus suggests that although the direction of the moral hazard argument is uncontroversial, the magnitude of this effect is too small for it to warrant a prominent role in the public policy debate. The clear economic benefits of economic stabilization and recovery in such economies as those of Mexico, Korea, and Brazil should thus weigh much more heavily in favor of a positive interpretation of the emergency support policies than should the moral hazard argument as the basis for a negative interpretation.[12]

10.   Rules versus Discretion

As with past experience in other areas of economic policy, especially monetary and exchange rate, the debate on how to involve the private sector in crisis resolution appears to have evolved toward a debate on whether rules fixed in advance or case-by-case discretion should determine policy. U.S. authorities appear to have emphasized the need for case-by-case treatment, whereas many of those in Europe appear to have been inclined more toward ex ante rules.

In terms of the framework above, the case for predetermined rules would presumably have to rest heavily on the notion that official support causes important moral hazard, so that private lenders should be set on notice in advance that if a country encounters difficulties, then this type of lending should be subject to a standstill, that type of lending to an exchange of instruments, and so forth, as the precondition for any official support. This approach tends to assume that the resulting economic costs of delayed return by the country to voluntary capital markets will be small, and that the benefits of avoiding moral hazard will be large.

The case-by-case approach, instead, is consistent with the need to determine a country's position along the Bagehot curve individually in each separate instance, and the need to deal with the different configuration of each country's debt structure in correspondingly different ways. Experience would suggest that the case-by-case approach is the more feasible of the two. If applied in a manner that

---

12   Finally, it should be reiterated that whereas moral hazard from official support has not been the dominant influence in most emerging markets, the special influence of geopolitical moral hazard appears to have played a role in excessive lending to Russia.

seeks to achieve the most voluntary possible private-sector involvement, the case-by-case approach also seems the more likely to secure favorable medium-term growth and stability in emerging market economies, by maximizing the chances of early return to market access.

Indeed, in contrast to fixed rules, creative ambiguity would seem the appropriate official-sector strategy. In particular, any firm pronouncement that the official sector will no longer provide large support packages is likely to be seen by the market as not credible with respect to countries large enough to have systemic consequences. In such cases, G-7 officials are to a considerable degree acting in their own countries' direct interests in avoiding international economic crisis. Yet it would be even more inadvisable for the public sector to state in advance that large economies with potentially systemic impact will be eligible for large emergency official support programs, as this would directly confirm too-big-to-fail moral hazard.

11.   Recent Patterns in Debt Workouts

The first half of 1999 saw the successful implementation of the Brazilian adjustment program with large international support and effective private-sector participation through voluntary maintenance of short-term bank credit lines. In contrast, in the rest of 1999 and in early 2000 there was an emerging pattern of default workouts among smaller economies (and one large one, Russia). It is useful to review these cases to distill the lessons they may have to offer for crisis resolution.

It is first useful, however, to note that there appears to have been an escalation in official-sector pressure for private-sector involvement during this period, and a strategy of using the smaller countries as test cases for pursuing less voluntary approaches toward this end. Early in 1999 the Paris Club informed Pakistan that it should achieve private-sector comparability with a debt rescheduling, and this eventually led to an exchange offer for Eurobonds late in the year. In the first half of 1999, there appears to have been similar official-sector signaling to Ecuador that, if it achieved its policy adjustments, it could appropriately call for the rescheduling of Brady bonds and Eurobonds. IMF management approval of a program for Ecuador in September represented a de facto endorsement of the default on Brady bonds adopted shortly before. In Romania, in September 1999 the Fund suspended its lending program after only 2 months primarily because the government had raised only $200 million from the private sector out of a $450 million burden-sharing target.

These economies of course differed from Mexico, Korea, and Brazil in more than size. Their underlying political-economic coherence was typically much weaker. They were essentially considerably further to the right along the Bagehot curve. Nonetheless, there is a case to be made that Ecuador, for example, could have pursued a different approach that placed much more emphasis on restoring confidence and maintaining market access. Indeed, Ecuador's subsequent finance minister declared the September 1999 default "disastrous."

Some of the principal patterns that seem to be emerging from the recent smaller-country workout cases include the following:

Communication with bondholders is not a major problem—A dominant concern of many in the discussions on international financial architecture in recent years has been that the rise in bond relative to bank financing could lead to chaotic conditions precluding orderly rescheduling in the future, because of the difficulty of communicating with, and organizing responses from, large numbers of small individual bondholders in contrast to the easier centralization of negotiations with large advisory-committee banks representing bank creditors. The recent workout cases suggest that these concerns are excessive, in large part because today's communication technology is vastly superior to that in the 1930s when protracted bond defaults reflected in part such problems.

In the case of the Ukraine, external sector difficulties (in part from higher oil prices) prompted suspension of payments to private and bilateral creditors in March 2000. The government sought a quick resolution consistent with its ongoing efforts to avoid default, and it chose the route of issuing an "exchange offer." Some $2.6 billion in debt of banks and bondholders would be replaced by 10-year bonds in euros and dollars at interest rates of 10 and 11 percent, respectively. The government would place $0.3 billion due in interest this year into escrow, to be paid out to creditors accepting the exchange.

About $1 billion of the bonds outstanding were held by "retail" holders, primarily in Europe. Most of the rest were held by hedge funds. Communication by e-mail through the Internet made it possible to enter into contact with approximately 25,000 holders. The government sent them its projections of debt payments and other economic materials. The exchange agreement was successfully completed within only three weeks.

Present-value preservation instead of major forgiveness facilitates exchange—One reason the Ukraine exchange was completed so quickly was that it did not seek debt forgiveness. Similarly, in November 1999 Pakistan offered to exchange at full face value about $600 million in Eurobonds coming due during the next two years. The government offered a dollar-denominated exchange bond paying

10 percent, due in four equal installments beginning December 2002 through December 2005. By the closing date in late December, more than 95 percent of bondholders had accepted the exchange. Pakistan had already obtained an agreement in July 1999 from eight commercial banks to maintain short-term trade credits through a trade facility paying 1.5 percent above LIBOR and with repayments during 2001 through mid-2002.

London Club arrangements have evolved with debt structure—In February 2000, Russia concluded a restructuring of $28 billion in previously rescheduled Soviet-era debt to commercial banks. The PRIN and IAN bonds that had been issued in December 1997 in exchange for pre-1992 bank claims were swapped at a discount of about 35 percent for 30-year Eurobonds with 7 years grace and front-loaded interest reduction.

The traditional advisory-bank structure of London Club arrangements in the 1980s was extended to include a number of investment banks, in light of the significant holdings of the PRIN and IAN bonds (close to one-third) in the financial markets rather than directly by banks.

The debt forgiveness involved, the new opportunity for a decisive resolution given the new president, and a sense that with strong oil prices the time to strike the best deal possible was at hand, seem likely factors in explaining the relatively lengthy hiatus between the August 1998 default and the early-2000 restructuring. Some might argue, however, that the London Club structure also played a role because it tends toward lengthy negotiations. This might be so, but that is not necessarily a bad thing for capital markets and emerging market economies. If a somewhat longer negotiation yields a more balanced outcome, in the future lenders to are more likely to have confidence that if things go wrong there will be a reasonable rather than highly disadvantageous outcome, and hence be more willing to lend.[13]

Consultation and negotiation with bondholders needs to be improved—At least some of the recent debt exchange offers do appear to have been deficient in mechanisms for broad dialogue and negotiation between the country and bond-

---

[13]   As shown by the historical experience of Mexico's precedent-setting Brady deal in 1989, in which Mexican authorities accepted a more moderate debt reduction— 35 percent—than even the Managing Director of the IMF had appeared to call for. This helped establish a sense of fair play that no doubt contributed to the revival of lending flows in the early 1990s. See Cline (1995: 220).

holders. Although in some cases leading bondholders have reportedly been consulted, the basic dynamic appears to have been presentation by the government of an offer on a take-it-or-leave-it basis with a tight response deadline. Although the terms of these deals typically have been set so as to leave the bondholders with a somewhat higher price than the secondary market price of the old bond prior to the exchange, there remains a considerable way to go in establishing greater bargaining symmetry in exchange offers. Moreover, only time will tell whether some of these deals will prove vulnerable to subsequent litigation in part as a consequence of their unilateral nature. Although bondholder negotiating symmetry thus appears to remain a problem in unilateral exchange offers, as discussed below mandatory collective action clauses in bond contracts would be a counterproductive solution. A more promising alternative might be clear expectation in IMF conditionality that substantial consultation with bondholders should take place in any exchange offers.

12.   Bond Clauses

Numerous proposals for reform of the international financial architecture call for the inclusion of collective action clauses and provisions for a bondholders' committee in the contracts for new bonds (e.g., G-22 1998). Portes (2000: 11–15) considers these "essential to create a viable alternative to ever-bigger bail-outs." He recognizes the central argument against them: that they would raise borrowing costs by making default easier for debtors, and thereby "seriously weaken the bonding role of debt," discouraging lending. He counters, however, that they would be good for creditors as well as debtors, because they would avoid unnecessarily prolonged disruptions in debt service and country growth. He cites domestic corporate bankruptcy experience as supporting evidence. The central problem with applying such experience, however, is that in domestic bankruptcies, physical collateral can be seized. This is not the case in sovereign lending, in part because of sovereign immunity. Portes recognizes this key difference but lightly dismisses it as "an empirical question."

There is a case for including clauses for collective action and bondholder representation within bond contracts on a *voluntary* basis (IIF 1999a: 99). Some borrowers may find it attractive to pay some premium to obtain the extra resulting flexibility. The central point is that if the official community were to *mandate* such clauses in all new emerging market bonds (for example, as a regulatory requirement for their flotation in all G-7 markets), there would be a strong signal to the capital markets that a new environment had arrived in which greater ease of rescheduling would be required rather than a (priced) market option.[14]

Eichengreen and Mody (2000) have conducted empirical tests contrasting spreads on bonds floated in the U.S. and U.K. markets to probe whether collective action clauses would in fact raise the cost of capital to emerging markets. Bonds issued in the United States require unanimous bondholder consent for restructuring, whereas those issued in the United Kingdom can be restructured with consent of a large majority (typically two-thirds or three-fourths). Such an empirical test is not really possible, however, if the policy recommendation in question is a uniform mandatory regime. Spreads observed in the past are largely irrelevant to this issue, because heretofore investors have not been particularly sensitized to these clauses. In contrast, they most certainly would be in a new future period of mandatory inclusion adopted for emerging markets bonds. It is all the more striking, therefore, that Eichengreen and Mody find issuers from countries with low credit ratings do indeed pay higher spreads (by about 150 basis points) on bonds issued in the U.K. than in the United States. Conversely, spreads are somewhat lower in the U.K.-issued bonds (by about 50 basis points) for highly rated countries; but the emerging market economies fall primarily into the lower-rated category.

As for the attribute of facilitating negotiations, such clauses did not in fact lead to this outcome in the cases of Pakistan and Ukraine, even though both had floated the Eurobonds in question under U.K. law. Instead, both countries issued unilateral exchange offers, with a short deadline for holders to accept or reject. Buchheit (2000) argues that sovereign debtors may be reluctant to use existing collective action clauses for formal negotiations, because they fear that calling bondholders into the negotiation process may facilitate their communication and thereby the mobilization of the threshold percentage (typically 25 percent) required to demand acceleration.

---

[14] Dilution of an emerging markets stigma by G-7 agreement to include collective action clauses in their own bonds cannot be considered a serious proposal, even though Canada and perhaps another country or two might be willing to do so. The implicit admission of potential default is not one most finance ministers are likely to make, and some G-7 countries with high debt (Italy, Japan) would be ill-advised to advertise this possibility.

13.   Stay of Litigation

Former IMF Managing Director Michel Camdessus (1999) has suggested that
Article VIII (2b) of the IMF's articles of agreement could be amended to provide
international official sanction of a temporary stay of litigation by creditors, to
"ensure that debtors have time to reach orderly resolution" in extreme situations.
This type of change, however, is precisely the type of international signal that
would likely have the effect of undermining confidence in capital markets about
the underlying balance between creditor and debtor that needs to be present if investors are to engage in sovereign lending. Potential lenders would have to be
concerned that such a provision would be applied in a political manner that
sought to shift more of the bargaining power in debt workouts to the debtor and
away from the creditors. It is the type of proposal that does not take into account
the induced, or general equilibrium, adverse effects predictable under debt theory
(e.g., Eaton and Gersovitz 1981) while attempting to solve a partial equilibrium
problem in a special case. Such measures would, moreover, erode the ability of
all emerging market sovereigns to precommit to nondefault, inflicting general
spillover costs (higher spreads, lower lending volumes) even if they conveyed
benefits to a specific country in default.

If the experience of the last decade had generated several significant cases in
which creditor litigation had caused economies to fall into deeper economic collapse, it might be appropriate public policy to at least weigh the potential de-
faulter benefits of a stay of litigation amendment against the likely more generalized debtor and creditor losses from capital market distortion through contract
erosion. However, that has not been the case. Unless and until litigation disruptions do become a clear problem, this would seem to be a proposal that should be
rejected.

14.   Bankruptcy Codes

Potentially the most extreme form of official intervention to force greater allocation of zero-sum losses from sovereign debtors to private creditors would be the
adoption of some form of international bankruptcy code for sovereign debt. The
G-10 (1996) considered and rejected this approach in its earlier review of financial architecture following the Mexico crisis. It noted:

> [Domestic bankruptcy procedures] reduce uncertainty [and] ... provid[e] the
> debtor with temporary protection from its creditors and access to interim fi-
> nance with some seniority ... enabl[ing] an enterprise whose value as a going

concern exceeds its break-up value to continue to operate.... Despite their advantages in a national context and some theoretical appeal by analogy, formal insolvency procedures do not appear to be either appropriate or feasible.... as a means of dealing with sovereign debt problems.... The analogy does not apply in some crucial respects, as it would be neither appropriate nor possible to replace the authorities responsible ... with a "new management," or to take possession of a state's non-commercial property. The need for additional protection from creditors has not in the past been a serious problem.... [S]overeign debtors have few assets to seize and some of these benefit from sovereign immunities. Any international insolvency convention would inevitably involve a long and cumbersome negotiation process. (pp. 7–8)

The G-10's reasons for rejecting the international bankruptcy approach remain valid despite the intervening crises in the past four years. In terms of sovereign debt theory, a bankruptcy code would seriously undermine countries' ability to precommit to nondefault for the purpose of preserving their reputation and hence subsequent market access. Such a code would tend to cast an aura of legitimacy on sovereign default and inevitably leave the impression that default in the future would be easier. An international arrangement would tend to be perceived in markets as highly susceptible to political considerations, which typically would tend to lean toward granting relief to sovereign debtors rather than toward enforcing creditor contract rights. For these reasons, it would tend to erode capital market confidence in repayment, with the consequence of reducing lending flows and raising interest rate spreads.

15.   Conclusion

The international official community and the private sector dealt decisively and effectively with the most dangerous financial crises of the late 1990s: those of Mexico, Korea, and Brazil. The proper lesson to draw is that the measures developed in these experiences are important to preserve and build upon. These include foremost the IMF's Supplementary Reserve Facility, which permits much larger emergency finance than traditional quota levels. They also include voluntary mechanisms for private-sector participation in crisis resolution, notably an arrangement to maintain short-term credit lines (or, if necessary, convert them to somewhat longer instruments at close-to-market terms). The approach adopted in these cases was perfectly consistent with central banking principles of forceful intervention to stem the effect of a temporary panic on an otherwise solvent bank (Bagehot 1873). This basic approach should not be reserved to system-threatening cases, because the economic benefits it can convey would thereby be deprived to smaller countries, and ex ante asymmetry on this basis could encourage too-big-to-fail moral hazard for authorities of large emerging market economies. The determination of applicability of this approach should be based on the location of the country on the Bagehot curve—whether it is close to strong solvency or severe insolvency—not on its size.

Despite the great attention to moral hazard stemming from these and other crisis resolution instances, there is little if any evidence to support the notion that there has been an excessive amount of lending to emerging markets at unduly low interest rates as a consequence. The large losses in several of these crises will be in financial institutions' memories for some time to come. Formal econometric tests reject the view that the Mexico support program in particular caused moral-hazard excess lending to emerging markets subsequently. It is crucial that those who focus on moral hazard also take into account the economic losses for all parties, including emerging market economies, that could be the alternative to adept official sector intervention even if such intervention inevitably must contain some degree of moral hazard risk.

Private-sector involvement in resolving financial crises should be on as voluntary a basis as possible. International practices and the institutions developed should be cognizant of the underlying theory of sovereign debt, whereby it is in a country's interest to preserve its reputation in order to borrow again in the future. Arrangements that erode this underlying "bonding" commitment will increase lender risk perception, tending to reduce capital flows and raise their cost. At the turn of the century, credit flows to emerging markets remain relatively fragile, especially through the vehicle of commercial bank loans. It is thus especially important that any modifications in the international financial architecture carefully take into account induced effects on emerging capital markets. Otherwise economic growth in these economies could be held below its potential in the coming decades.

Bibliography

Bagehot, W. ([1873] 1917). *Lombard Street.* 14th ed. John Murray. London: Kegan, Paul & Co.

Buchheit, L.C. (2000). Sovereign Debtors and Their Bondholders. Testimony before the Meltzer Commission. Available on http://phantom-x.gsia.cmu.edu/IFIAC.

Camdessus, M. (1999). From the Crisis of the 1990s to the New Millenium. Remarks to International Graduate School of Management (IESE), 27 November. Available on www.imf.org.

Cline, W.R. (1995). *International Debt Reexamined.* Washington: Institute for International Economics.

Cline, W.R. (1998). IMF-Supported Adjustment Programs in the East Asian Financial Crisis. IIF Research Paper 98-1. Institute of International Finance, Washington.

Council on Foreign Relations (1999). *Safeguarding Prosperity in a Global Financial System: the Future International Financial Architecture.* Report of an Independent Task Force: Carla A. Hills and Peter G. Peterson, Co-Chairs; Morris Goldstein, Project Director. Washington: Institute for International Economics, for Council on Foreign Relations.

Eaton, J., and M. Gersovitz (1981). Debt with Potential Repudiation: Theoretical and Empirical Analysis. *Review of Economic Studies* 48 (April): 284–309.

Eichengreen, B., and A. Mody (2000). Would Collective Action Clauses Raise Borrowing Costs? NBER Working Paper 7458, Cambridge, Mass.

IIF (1999a). *Report of the Working Group on Financial Crises in Emerging Markets.* Washington: Institute of International Finance.

IIF (1999b). *Data Release Practices of Emerging Market Economies: 1999 Assessment.* Washington: Institute of International Finance.

IIF (2000a). *Capital Flows to Emerging Market Economies.* Washington: Institute of International Finance. Available on www.iif.com.

IIF (2000b). *Comparative Statistics for Emerging Market Economies.* Washington: Institute of International Finance.

G-10 (1996). *The Resolution of Sovereign Liquidity Crises*, A Report to the Ministers and Governors prepared under the auspices of the Deputies. Washington, D.C.: International Monetary Fund.

G-22 (1998). *Report of the Working Group on International Financial Crises.* Washington: Group of 22. Available on www.worldbank.org.

Meltzer Commission (2000). *Report of the International Financial Institution Advisory Commission ("Meltzer Commission").* Washington: IFIAC. Available on http://phantom-x.gsia.cmu.edu/IFIAC.

Portes, R. (2000). Sovereign Debt Restructuring: the Role of Institutions for Collective Action. Paper presented at World Bank-IMF-Brookings Institution conference on Emerging Markets in the New Financial System: Managing Financial and Corporate Distress, March 30–April 1, Florham Park, New Jersey.

Stiglitz, J. (2000). The Insider. *New Republic*, 17 April.

Zhang, X.A. (1999). Testing for "Moral Hazard" in Emerging Markets Lending. IIF Research Paper 98-1. Institute of International Finance, Washington.

F72 016
F34 019

p55: # Comment on William R. Cline

Theo S. Eicher

Bill Cline provides a detailed survey of key aspects of financial crisis management and he proposes interesting remedies. The paper is compelling to read because he peppers his theoretical points with concrete, real world examples. In my comments I constrain myself to general points regarding the recommendations that he proposes and to some specific questions that concern the remedies he puts forth.

In Cline's view, the management of financial crises falls into three categories: (a) macroeconomic policies, (b) international official support, and (c) the nature of private-sector involvement in debt renegotiations. Dealing with all three areas in one paper is a magnificent task, so he specializes and focuses on private-sector participation and international support.

I would like to add one category that I think has not received the attention it deserves. I am surprised how often social measures to distribute the burden of the crisis are not fully considered when economists discuss crisis management. I am not mentioning this as a moral imperative, but for purely positive economic reasons. The advances in the formal literature on the political support in the past decade highlight how crucial distributive measures are to the welfare of the economy and to the survival of governments and decision makers.

Let me now examine in detail the areas Cline emphasizes. First are specific country policies; what should these look like? Cline emphasizes nicely the diversity of financial crises in the 1990s. This discussion might be augmented with common themes in all countries, such as how to deal with contagion, common shocks to commodity prices, relative price and demand shocks transferred via trade linkages, and competitive devaluations. Cline focuses firmly on the private sector and constrains himself to arguing solely for a forceful adjustment program, for reforms of the financial sector (if fiscal balance sheet is clean), and for tight monetary policy and high interest rates.

The discussion on the voluntary vs. involuntary participation of the private sector in debt negotiations is very interesting. After presenting detailed evidence, he concludes that negotiations should be as voluntary as possible. This fits well with informational asymmetries stories, as transparency, true reforms, good data availability, and clear adjustment programs all mitigate information costs and the effects of such asymmetries on investors.

His specific proposals relating to the magnitude of official support are most intriguing. Cline is confident that "public sector bailouts have not caused subsequently excessive private sector lending to emerging markets" (p. 68) and he postulates that moral hazard is not a crucial problem that requires serious priority in policy design. My comments on this view of the world are twofold. First, even if it were true that moral hazard is not prevalent months after the crises, it does not imply that the results of informational asymmetries will not return to contribute to future crises. Second, in my reading of the literature excessively risky lending caused in large part by moral hazard and adverse selection contributed greatly to the occurrence and size of most financial crises in the 1990s. Cline's paper, however, highlights nicely how important serious empirical documentation of moral hazard is—and that, aside from the paper he cites, there exists no systematic analysis of the phenomenon.

This brings me to the concept of the Bagehot curve, the idea of providing liquidity without limit to solvent institutions (Bagehot's version), or to solvent countries (Cline's version). Alternatively, if the institution or country is insolvent, not a penny should be forthcoming to alleviate a crisis. For countries that exhibit some degree of solvency, Cline proposes some degree of forthcoming liquidity, according to the shape of his Bagehot curve. I find this idea intriguing, but think it needs to be fleshed out more. "Solvency" sounds like such a simple concept, but what measure(s) of solvency should be employed to evaluate countries? Can there be agreement on any one measure of solvency, and who is to make this decision? Should solvency be the only measure? Is it realistic to abstract from political considerations and suggest that insolvent countries should be denied liquidity? Who should assess country solvency?

Finally, Bagehot's idea works only if moral hazard is ruled out, or if the effects of moral hazard can be neglected. But for economists, moral hazard and adverse selection are textbook outcomes of informational asymmetries. This implies Cline has to argue that informational asymmetries or their effects are negligible.

This leads me to a discussion of the significance of the debt theory that Cline indicates is the basis of his analysis. He cites Eaton and Gersovitz (1981), where the probability of default raises the interest rate and reduces the loan amount. This relationship is again generated via informational asymmetries and their effects, as lenders must feel the borrower will make every effort to repay, otherwise they demand a risk premium, or cease lending. It is interesting how much weight Cline attributes to these theoretical foundations given that in his own mind, moral hazard, the direct outcome of informational asymmetry, has been exaggerated.

Debt theory provides guidance for crisis management because it provides a framework with which to examine the causes of a crisis. How we perceive these causes depends on our analytical framework. The framework that economists commonly employ is linear, in the sense that our models have linear transition paths as the economy experiences changes in the underlying parameters. This simplifies the math but more importantly, it simplifies our outlook of the world.

There is ample evidence that initial positive results due to financial liberalization are often wiped out in developing countries due to capital flow reversals (see Klein and Olivei 1999; Reisen and Soto 2001; Edwards in this volume). In a linear world, this implies that the country initially engaged in good policy (i.e., financial liberalization), but that bad polices at a later date wipe out all the gains.

Nonlinear transitions result when Eaton and Gersovitz is integrated into a semiendogenous growth model, where the interest rate rises in the country's debt to equity ratio (as confirmed by Edwards 1984). The transition due to financial liberalization with excessively low costs for foreign funds is nonlinear in that the country first booms, then it experiences a shortage of foreign capital and a rising debt cost and finally capital outflows and recession (such dynamics are supported by data, see Eicher et al. [2000]). This implies that crises may already be programmed in, and hence crisis management must be much more carefully planned, depending on whether the economy is in equilibrium or not.

Bibliography

Eaton, J., and M. Gersovitz (1981). Debt with Potential Repudiation: Theoretical and Empirical Analysis. *Review of Economic Studies* 48(2): 289–309.

Edwards, S. (1984). LDC Foreign Borrowing and Default Risk: An Empirical Investigation, 1976–1980. *American Economic Review* 74(4): 726–734.

Eicher, T.S., S.J. Turnovsky, and U. Walz (2000). Optimal Policy for Financial Market Liberalizations: Decentralization and Capital Flow Reversals. *German Economic Review* 1(1): 19–43.

Klein, M., and G. Olivei (1999). Capital Account Liberalization, Financial Depth and Economic Growth. NBER Working Paper 7384. Cambridge, Mass.

Reisen, H., and M. Soto (2001). Which Types of Capital Inflows Foster Developing-Country Growth? *International Finance* 4(1).

II.

The Changing Structure of Financial Markets

Leslie Hull and Linda L. Tesar

F31 F21
F32 019
016

(selected countries)

The Structure of International Capital Flows

1. Introduction

Over the course of the last decade, the volume of cross-border capital flows has expanded at a remarkable rate. Between 1990 and 1997, the volume of capital flows between industrialized countries (measured as the sum of gross inflows and outflows) doubled, reaching a peak of $4.5 trillion.[1] Net private capital flows to emerging markets nearly quadrupled, expanding from less than $50 billion in 1990 to a projected $190 billion in 2000.[2] To put these figures in some perspective, over the 1990–98 period, world GDP increased by roughly 30 percent and world trade by a little over 50 percent.[3] The increased volume of capital flows has been attributed to a number of factors: the improvement in communications technology, the deepening and liberalization of financial markets in many developing countries, increased demand on the part of investors for higher rates of return and greater opportunities for diversification, and the continuing privatization of state-owned enterprises. What is uncontested is that sizable cross-border capital flows are now a permanent fixture of the global economy.

The composition of international capital flows also changed markedly in the 1990s. Portfolio inflows to emerging markets expanded from a mere trickle in 1990 to a peak of $51 billion in 1993, accounting for over 50 percent of total capital inflows in that year.[4] External finance shifted from a reliance on official flows to borrowing on private markets. The share of foreign direct investment

---

[1] See IMF, *Yearbook Balance of Payments Statistics* (various issues).

[2] Forecast of the Institute of International Finance (2000). The forecast of the World Bank is less optimistic, with a projected net flow of $145.5 billion in 2000. Net private capital flows to all developing countries reached $257 billion in 2000.

[3] Data on world GDP and trade from IMF, *International Financial Statistics* (various issues). World trade is measured as the sum of world exports and world imports.

[4] Data from IFC, *Emerging Stock Markets Factbook* (1998).

(FDI) increased from less than 20 percent of total inflows in 1985 to over half of all inflows in 1998.

During the first wave of capital market liberalization in emerging markets, most academicians and policymakers shared the view that a globalized capital market would provide significant benefits. It was widely believed that increased financial market integration would facilitate the movement of capital from the capital-rich to the capital-poor and would help countries as a whole diversify risk. Increased competition would make financial intermediation more efficient and bring the global supply of savings to parts of the world previously left out of the expansion of the global economy.

The optimism of the early 1990s then came face to face with a bewildering series of financial market crises. In a litany that is now all-too-familiar, the Mexican peso crisis in late 1994 was followed by the Asian meltdown, with its spill-overs to Hong Kong, Russia, and Latin America. The real effects of these crises were substantial and the afflicted countries are only now beginning to recover. The crises have left us with a difficult set of questions. Were the crises triggered by fundamental macroeconomic conditions? Did capital market integration lay the groundwork for global instability? Were capital flows the source of the transmission of crises across countries? If capital markets were the root cause of the crises and their transmission, what is the appropriate policy response to prevent or at least mitigate future crises?

Before we can begin to answer questions about "what went wrong" in international capital markets, we need a framework to establish what capital flows look like when everything "goes right." Our paper is only a first step in this direction, but our intention is to develop an analytical framework for thinking about the determinants of the composition of international capital flows and to compare the broad predictions of the model with the data. In the next section, we begin with a simple closed-economy model of borrowing and lending with asymmetric information and costly monitoring. We then ask how this framework must be modified to think about the structure of capital flows in an open economy. In Section 3, we contrast the implications of this model with data on international capital flows in industrialized and developing countries. In light of the global experience with capital flows in the second half of the 1990s, we then reconsider how the basic model with asymmetric information would have to be extended in order to explain short-term fluctuations in capital flows and balance-of-payments crises.

2.   Corporate Finance in a Global Economy

A great deal of research has focused on the channels of capital flows under various assumptions about the completeness of financial markets, the information set available to market participants, and the regulations that distort private decisions. The approach typically adopted is to focus on one channel of international capital flows and examine the optimal policy response to distortions in that market.[5] Surprisingly little work has attempted to study the composition of capital flows in a framework that allows for multiple channels of capital flows.[6] If, as has been suggested, recent financial crises occurred as a consequence of the composition of capital flows to emerging markets, a model that allows for multiple channels of capital flows is necessary both to understand recent events and to develop an appropriate policy response.

As a small step in this direction, we characterize the composition of capital flows when private agents have access to bond markets, equity markets, and bank loans. We begin with the simple debt-equity decision in a closed economy with asymmetric information between borrowers and lenders. We extend the market structure to include bank loans when there is private information, costly bankruptcy and costly monitoring by banks. We then apply the model to three open-economy scenarios to derive some basic implications for the composition of capital flows. Our model falls short of a full equilibrium analysis in that it takes the supply of savings as given. The advantage of the model, however, is that it provides a rich description of the demand side of the market for corporate finance and can produce an endogenous decomposition of finance into debt, equity, and bank loans.

a.   The Pecking Order of Corporate Finance in a Closed Economy

Consider an economy populated by a large number of entrepreneurs endowed with identical amounts of wealth. Each entrepreneur has access to a project for

---

5   To name just a few examples, Corsepius et. al. (1989) and Razin et. al. (1998, 1999) study the incentives for FDI under asymmetric information. Other studies focus on problems of herding and contagion in equity markets (see for example, Calvo and Mendoza (2000) and Chari and Kehoe (1997)). Self-fulfilling liquidity crises in debt markets are examined in Rodrik and Velasco (2000) and Chang and Velasco (1999a, 1999b).

6   Exceptions include Calvo (1998), Hull and Tesar (2000), Lane and Milesi-Ferretti (2001), Montiel and Reinhart (1999), and Schnitzer (1995).

producing the final consumption good, Y. The outcome of the entrepreneur's project can take one of two values, Y_H or Y_L, where $Y_H > Y_L$. Entrepreneurs differ in their likelihood of producing Y_H. "Good" entrepreneurs have a probability of a high outcome, p^G, and "bad" entrepreneurs have a probability of a high output of p^B, where $p^G > p^B$. To produce the final good, entrepreneurs must make an investment, I, in the project which exceeds their endowment of wealth, w.[7] Thus, entrepreneurs must go to the capital market to finance the project. The question we ask is, what type of arrangement will be made between the agents in the economy who are endowed with wealth and the entrepreneurs who are endowed with projects? Should savers in the economy lend directly to the entrepreneur or invest indirectly through financial intermediaries? If they invest directly, should they offer a loan with a prespecified rate of repayment or should they buy ownership shares in the project?

If we assume there is no private information (i.e., investors can observe the entrepreneur's type, the amount of investment in the project, and the project's outcome), there is no potential adverse selection or moral hazard problem. If we also assume that there are no transactions costs in writing or enforcing contracts between borrowers and lenders and no distortionary taxes, then the entrepreneur's choice of financing policy is irrelevant. Given full information, perfect markets, and rational investors, Modigliani and Miller (1958) showed that the value of a firm depends only on the payoffs of the project. In a frictionless world, the distribution of the payoffs among various lenders and shareholders are priced correctly in the market and will therefore be irrelevant for the investment decision and therefore for the value of the firm.

We know, however, that there are a number of reasons why this irrelevance proposition does not hold. Taxes, bankruptcy costs, and private information will change the incentives facing both borrowers and lenders and will affect the optimal capital structure of the firm. The optimal capital structure, then, will be a function of the legal, informational, and technological frictions facing the entrepreneur and potential lenders.

Suppose, for example, that entrepreneurs know their type (that is, they know their probability of producing high output) but potential lenders do not. Entrepreneurs have two options for financing their project. One option is to issue a bond that commits them to an ex ante schedule for repaying the loan. If the entrepreneur fails to make payment, the bondholders can force the firm into bankruptcy and bondholders have priority claims to the project's assets over shareholders. The advantage of predetermined interest payments is that entrepreneurs, rather

---

[7]   We assume for now that there is no discretion over the amount invested. We return to the issue of endogenous investment and moral hazard below.

than lenders, bear the risk of a bad outcome and therefore "bad" firms will be less likely to mimic "good" firms in the bond market.[8]

The second option is to issue equity in the project that pays off a predetermined share of output. The disadvantage of equity is that it is a costly means of finance for "good" firms due to the asymmetry of information between entrepreneurs and lenders. From the perspective of a prospective shareholder, "good" firms are indistinguishable from "bad" firms and therefore the prospective shareholder will demand a premium to compensate for the risk of purchasing a "lemon."

Myers and Majluf (1984) demonstrated that in such an environment there is a ranking of the different types of corporate finance. Internal finance (which is taken as exogenous in our example) is at the top of the pecking order. When external finance is required, debt is preferred to equity. In a more complicated scenario like the one we consider next, firms with the highest credit rating will issue bonds first and then move down the pecking order to some mix of bank loans and equity.

b.   The Pecking Order with Costly Bankruptcy and Bank Monitoring

Myers' theory is useful for thinking about the trade-off between debt and equity with adverse selection but it is silent about the role for bank finance, an important channel of international capital flows. We next ask what happens to the pecking order when bank loans are added to the menu of options for corporate finance. Following Bolton and Freixas (2000),[9] we extend the model to two periods; we then assume that the probability of high output in the first period can take on a range of values across entrepreneurs and is fully observable to potential creditors. Probabilities in the second period, however, are private information to the entrepreneur.[10] The entrepreneur is now labeled "good" or "bad" depending on the outcome in the second period and we assume that there is no correlation between first and second period probabilities. This setup implies that some "good" entre-

---

[8] If lenders cannot observe the amount invested in the project, debt contracts will also help solve the moral hazard problem by shifting the risk of low output due to underinvestment back onto the borrower.

[9] We do not fully develop the model but instead provide a general description of the structure of the model and its main implications for international capital flows.

[10] We follow Bolton and Freixas (2000) in assuming that the probability of a successful project in the second period, p_2, can take on values of 0 or 1. We define v as the creditors' probability assessment that p_2 is equal to 1.

preneurs will have a low credit rating in the first period and will therefore face a higher cost of financing the project. Similarly, some "bad" entrepreneurs will have a high credit rating and will create an information-dilution (or lemons) problem in bond and equity markets.

The entrepreneur has three options for financing investment in the project. Firms can issue bonds with a prespecified sequence of repayments, $[R_1, R_2]$. If bondholders force the firm into bankruptcy, they have priority over shareholders. Firms may also issue equity. Even though there is still a lemons problem in the equity market, the advantage of issuing equity is that entrepreneurs that are perceived to be high-risk borrowers may be able to obtain financing even when banks or bondholders are unwilling to extend credit to the project.

The third type of finance is a bank loan. Continuing to follow Bolton and Freixas (2000), we assume that a bank contract specifies a stream of payments, $[R_1', R_2']$. However, if the firm defaults on its first-period payment, the bank can obtain information about the firm and renegotiate R_2'. In the event of a bankruptcy, bank loans are paid ahead of outstanding debt and equity claims.[11] Bank finance thus comes with the flexibility of restructuring and the possibility of information sharing between the entrepreneur and the bank. However, bank finance entails an intermediation or monitoring cost that must be covered by the spread between the interest payments received on its portfolio of loans and the interest paid to its depositors. We will take this spread as exogenous and examine the effect of decreases in the cost of monitoring on the distribution of external finance.

The model leads to a modified pecking order of capital finance that depends on the spread, ρ, and v, creditors' beliefs about the probability of Y_H in period 2.[12] The pecking order is illustrated in Figure 1. In general, firms with good credit ratings will self-select to raise capital through the bond market, avoiding both the "lemons problem" of the equity market and the bank's intermediation costs. Some firms with high credit ratings may also choose to issue equity along with bonds. This is because the "lemons" premium associated with issuing equity for relatively low-risk firms is preferable to the risk of an inefficient bankruptcy

---

11   Because of the priority of bank loans over debt in bankruptcy, there may be inefficient liquidation of the firm by banks relative to the full information equilibrium. For details, see Bolton and Freixas (2000).

12   To support an equilibrium, additional assumptions are needed restricting the relative magnitudes of high and low output, the resale value of the firm in bankruptcy, expected period 2 output, and the amount of external finance demanded by the firm. We follow Bolton and Freixas in assuming that the bank may securitize part of its portfolio of loans. We impose the assumptions needed to support an equilibrium in the numerical simulations below.

that could occur with bank finance.[13] Medium-risk firms that cannot raise funds on the bond market rely on bank loans and/or equity. Firms with poor credit ratings either receive no investment funds at all and therefore cannot undertake the project, or rely solely on the equity market.

Figure 1: The Pecking Order for Finance in a Closed Economy

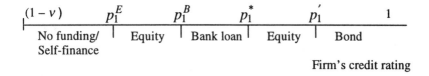

To gain some intuition about the magnitudes of the different types of corporate finance in equilibrium, we solve the model numerically under different assumptions about the distribution of risk across firms and the spread. The model is solved by finding the break-even points for each of the alternative forms of finance. Table 1 illustrates the allocations under a particular set of parameter values for high- and low-output, creditors' beliefs about the probability of high future output and the spread.[14] In the first column of the table we show the distribution of finance between bonds, equity, and bank loans and the credit rating thresholds when the expected probability of Y_H (v) is assumed to be high and banks face a spread of 10 percent. In this benchmark case, 78 percent of firms are able to obtain some external finance. Most entrepreneurs go to the bond market (56 percent), approximately 30 percent rely on bank loans and the remaining 15 percent issue equity. The model produces a counterfactually high debt-to-equity ratio relative to economy-wide debt-to-equity ratios in industrialized countries. However, the model is still useful for comparative statics exercises.

A reduction in the probability of high output in period 2 (second column of Table 1) results in an upward shift of the cutoff points for each type of finance. Roughly 25 percent are unable to raise external finance and of those that do; more must rely on bond financing. The intuition for the increase in bond finance is that as expectation of higher future output declines, the lemons problem in the

---

[13] Because banks are given priority over bondholders in bankruptcy, the bank may force some firms into bankruptcy that in a full-information world would be allowed to undertake the project in period 2.

[14] The model is solved in GAUSS. The program can be obtained from the authors upon request. Our benchmark parameters are somewhat arbitrary. We leave the challenge of a full calibration of the model for future research.

Table 1: Numerical Results

	Benchmark	Low probability of Y_H	High cost of financial intermediation	Developing country (2) & (3)
Parameters				
E (p_2)	0.90	0.80	0.90	0.80
Spread (ρ)	0.10	0.10	0.20	0.20
High output (Y_H)	0.60	0.60	0.60	0.60
Low output (Y_L)	0.35	0.35	0.35	0.35
Expected output	0.58	0.55	0.58	0.55
Minimum thresholds				
Equity (p_1^E)	0.30	0.40	0.30	0.40
Bank loan (p_1^B)	0.32	0.43	0.37	0.47
Equity 2(p_1^*)	0.53	0.59	0.54	0.60
Bonds (p_1')	0.61	0.65	0.61	0.65
Percent receiving external finance	77.78	75.00	77.78	75.00
Composition of external finance				
Equity	14.83	14.96	19.50	19.86
Bonds	55.56	58.83	55.56	58.82
Bank loans	29.61	26.22	24.94	21.32

equity market increases and firms must bear more of the risk of a low outcome. An increase in the cost of intermediation (represented by an increase in the spread in column three) leaves the proportion of external financing and the share of bonds in external financing unchanged. However, the change in the spread increases the relative share of equity finance to bank loans.

The fourth column shows the impact of a change in both the spread and expected period-2 output. One interpretation of the numerical example is that developing countries are likely to face *both* a higher cost of financial intermediation and a lower expected fraction of good firms. The joint effect relative to the benchmark case is that fewer firms receive external finance, bond finance is larger due to the lemons problem, and the share of bank finance is lower due to the higher spread.

c.   The Composition of Borrowing in an Open Economy

Thus far in the analysis, we have examined the allocation of a fixed supply of savings (which determined the spread) across firms. As an economy opens to global financial markets, both the opportunities for financing domestic projects *and* the supply of savings are likely to change. The joint effect of openness on the incentives for saving and investment will determine the country's net demand for finance and its composition.

As a starting point for thinking about this problem, we examine three special cases. The first case characterizes the impact of capital market integration in two identical economies. The second describes the impact of capital market integration on a small open economy that takes the world cost of capital as given. The third scenario considers capital market integration when foreign lenders are at an information disadvantage relative to domestic lenders. These three cases will serve as useful benchmarks for evaluating the data on international capital flows in Section 3.

Case 1: Diversification of Risk

First, consider two economies that are identical in all respects except that in each country, some fraction of project risk is country-specific. Now, all lenders—banks, bondholders, and shareholders—can diversify their portfolios across home and foreign entrepreneurs. Assuming there is no additional information asymmetry between domestic entrepreneurs and foreign lenders, the benefits of diversification should decrease the risk premium associated with investment, lower the spread, and raise the total supply of savings.

To see this, consider the problem facing a risk-averse lender. The value of the return on any investment, R, be it in the form of a bond, equity, or a bank loan, will be evaluated according to the asset-pricing equation of the consumption-CAPM:

$$q_t = E_t\,[u'(C_{t+1})\,R_{t+1}]\,/\,u'(C_t)$$
$$= E_t\,[u'(C_{t+1})/u'(C_t)]\,E_t\,[R_{t+1}] + [1\,/\,u'(C_t)]\,cov_t[\,u'(C_{t+1}),\,R_{t+1}\,],$$

where C denotes the lender's consumption, $u(C)$ is a utility function with the usual properties, E_t is the expectations operator conditional on date t information, and primes denote first-derivatives. If the lender is risk-averse, the more diversified the lender's portfolio, the lower the covariance between the expected marginal utility of future consumption and the payoff on any given asset. Therefore, the lender will demand a smaller risk premium per unit of investment, the cost of capital will fall, and the amount of investment in the economy will rise.

Given the symmetry between the two countries, the only incentive for capital flows is the diversification of portfolios on the part of lenders. The composition of capital inflows and capital outflows will therefore be identical and will mirror the composition of borrowing within the domestic economy. The (ex ante) volume of net flows will be zero.[15] One possible application of this case is capital flows between industrialized countries, where the economies have fairly similar risks, information asymmetries between domestic and foreign lenders are relatively small, and property rights are protected by international law. The model predicts that gross capital inflows and gross capital outflows will be of roughly the same magnitude and share the same composition.

Case 2: Small Country with Firms with Low Credit Ratings

In the second scenario, consider a country that is populated with firms that tend to have higher risk relative to the global population and therefore lower credit ratings. The simplest way to think about the implications for international capital flows is by looking at Figure 1. Since most firms are located in the lower end of the pecking order, the model implies that the composition of capital flows will be skewed toward equity and bank loans. Note also that because domestic firms are riskier than the global population, capital flows will be more volatile than flows to countries with less risky firms.

Case 3: International Information Asymmetry

Suppose that the domestic and foreign countries are identical in all respects, except that foreign lenders are at an informational disadvantage relative to domestic lenders. This can be modeled in a number of ways, such as an increase in the perceived variance associated with foreign projects or an increase in the cost of monitoring the foreign entrepreneur. Razin et al. (1998) make the extreme assumption that domestic lenders know the borrower's type with certainty, while foreign lenders do not.[16] In their setup, foreign direct investment is assumed to

---

[15] There could be current account imbalances ex post depending on the outcomes of the projects in the two countries. If the number of firms in each country is infinite, so that complete risk sharing is possible, then current accounts may also be zero ex post.

[16] Razin et al. (1998) assume that investment decisions are made prior to the financing decision; therefore, the nature of the financing contract will have an effect on the value of the firm and economic welfare. We will return to the problem of endogenous investment and moral hazard in Section 4 below.

fully circumvent the information asymmetry—by taking a controlling interest in the firm, FDI essentially becomes an international form of self-finance. The empirical implication is that in countries where information asymmetries are particularly important, FDI dominates all other forms of finance in the pecking order for global capital.

3. International Capital Flows in Industrialized and Developing Countries

In this section we discuss recent trends in international capital flows in industrialized and developing countries. We examine whether there is evidence of a "pecking order of finance" in international data.

a.   Industrialized Countries

Table 2 shows gross inflows, gross outflows, and net inflows in industrialized countries over the 1988–1998 period. Total flows are disaggregated into direct investment, portfolio investment (equity, debt, and money market and financial derivatives), and loans and other liabilities. Several patterns emerge from the data. First, net inflows are only a small fraction of gross inflows—indeed, over the last decade net inflows accounted for an average of only one percent of gross inflows. In other words, most of the capital that originates in industrialized countries is reinvested in industrialized countries. Second, there is considerable volatility in capital flows over time at the aggregate level as well as in the components of the capital account. The table illustrates the dramatic increase in gross inflows between 1994 and 1997 fueled primarily by the run-up in debt in the form of bank loans and bonds.

The composition of capital flows is shown in Table 3. Comparing the average shares in the last two columns of Table 3, it appears that the shares of direct investment and bonds have been fairly constant over time. The one relatively small shift in the composition of capital inflows and outflows has been the decline in bank lending offset by a five to six percent increase in the share of equity finance. The data suggest that bonds are the primary source of international finance, followed by bank loans and direct investment.

The similarity in both the magnitudes of capital inflows and outflows and their decomposition is consistent with our discussion of "Case 1," in which capital flows are primarily driven by international diversification and perhaps marginal

Table 2: Capital Flows in Industrialized Countries, 1988–1998 (billions of dollars)

		1988	1989	1990	1991	1992	1993	1994	1995	1996	1997	1998
Inflows	Direct investment	132.1	166.1	169.6	112.9	117.7	136.5	139.5	208.9	222.5	272.1	458.3
	Reinvested earnings	11.7	8.0	-7.9	-16.0	-13.1	1.5	17.5	39.3	35.4	50.1	39.6
	Other direct investment	120.3	158.1	177.5	128.9	130.8	135.0	122.0	169.3	187.1	220.3	418.7
	Portfolio investment	216.8	350.5	213.6	410.9	385.3	613.4	316.2	541.5	775.6	855.6	829.9
	Equity	34.0	75.4	-7.6	99.4	90.0	181.9	109.9	120.0	173.1	208.6	315.7
	Debt	173.3	268.9	208.5	306.7	302.1	441.1	155.0	370.1	565.6	577.2	434.2
	Money market & fin der	9.5	6.2	12.7	4.8	-6.7	-9.6	51.2	51.3	36.9	69.7	80.0
	Other liabilities	534.0	675.0	473.0	-57.4	177.5	123.0	274.4	209.9	436.4	749.1	310.1
	Loans	113.3	187.6	200.7	87.2	106.0	224.5	-6.7	304.2	240.9	382.7	186.5
	Other	420.6	487.4	272.3	-144.6	71.5	-101.5	281.1	-94.3	195.5	366.4	123.6
	Reserve liabilities											
	Total	882.8	1191.5	856.2	466.4	680.5	872.9	730.1	960.3	1675.5	2259.5	1784.9
Outflows	Direct investment	-161.6	-211.0	-224.4	-186.7	-178.8	-206.5	-211.4	-278.3	-339.7	-411.2	-585.5
	Reinvested earnings	-35.4	-39.4	-45.2	-33.6	-26.6	-48.3	-66.5	-96.5	-110.0	-126.8	-103.3
	Other direct investment	-126.2	-171.7	-179.3	-153.0	-152.1	-158.2	-144.8	-181.7	-228.7	-284.4	-482.2
	Portfolio investment	-207.1	-276.8	-169.1	-317.0	-328.3	-537.7	-306.8	-398.4	-576.1	-650.3	-896.6
	Equity securities	-28.8	-83.1	-25.3	-90.6	-76.9	-153.8	-125.9	-100.6	-153.1	-188.7	-309.3
	Bonds and notes	-178.0	-193.3	-141.4	-212.1	-242.1	-357.5	-162.7	-289.3	-413.5	-415.6	-505.5
	Money market & fin der	-4.1	-3.7	-2.4	-14.3	-9.3	-26.5	-18.2	-8.5	-9.5	-46.0	-81.8
	Other assets	-408.0	-616.2	-325.0	-44.2	-160.7	-209.0	-65.5	-232.1	-512.3	-735.4	-195.8
	Loans	-93.1	-177.2	-184.7	45.7	-66.1	-153.3	-79.7	-304.4	-115.9	-399.2	-96.4
	Other	-314.9	-438.9	-140.3	-89.9	-94.6	-55.7	14.2	72.3	-396.4	-336.2	-99.4
	Reserve Assets	-38.8	-26.7	-57.5	-14.1	3.1	-18.9	-35.6	-80.4	-77.2	-24.4	19.3
	Total	-815.5	-1131	-776	-562	-664.7	-972.1	-619.3	-989.2	-1621	-2220	-1755
Net inflows		67.3	60.9	80.2	-95.6	15.8	-99.2	110.8	-28.9	54.3	39.1	29.9

Note: "fin der" denotes financial derivatives.

Source: IMF, Balance of Payments Statistics Yearbook (various issues).

Table 3: Industrialized Countries Shares of Total Flows (in percent)

	1988	1989	1990	1991	1992	1993	1994	1995	1996	1997	1998	Average	
												1988-92	1993-98
Inflows													
Direct investment	15.0	13.9	19.8	24.2	17.3	15.6	19.1	21.8	13.3	12.0	25.7	18.0	17.9
Portfolio investment	24.6	29.4	24.9	88.1	56.6	70.3	43.3	56.4	46.3	37.9	46.5	44.7	50.1
Equities	*3.9*	*6.3*	*-0.9*	*21.3*	*13.2*	*20.8*	*15.1*	*12.5*	*10.3*	*9.2*	*17.7*	*8.8*	*14.3*
Bonds & money market	*20.7*	*23.1*	*25.8*	*66.8*	*43.4*	*49.4*	*28.2*	*43.9*	*36.0*	*28.6*	*28.8*	*36.0*	*35.8*
Loans & other liabilities	60.5	56.7	55.2	-12.3	26.1	14.1	37.6	21.9	26.0	33.2	17.4	37.2	25.0
Outflows													
Direct investment	19.8	18.7	28.9	33.2	26.9	21.2	34.1	28.1	21.0	18.5	33.4	25.5	26.1
Portfolio investment	25.4	24.5	21.8	56.4	49.4	55.3	49.5	40.3	35.5	29.3	51.1	35.5	43.5
Equities	*3.5*	*7.4*	*3.3*	*16.1*	*11.6*	*15.8*	*20.3*	*10.2*	*9.4*	*8.5*	*17.6*	*8.4*	*13.6*
Bonds & money market	*21.8*	*17.1*	*18.2*	*37.7*	*36.4*	*36.8*	*26.3*	*29.2*	*25.5*	*18.7*	*28.8*	*26.3*	*27.6*
Loans and other assets	50.0	54.5	41.9	7.9	24.2	21.5	10.6	23.5	31.6	33.1	11.2	35.7	21.9
Reserve assets	4.8	2.4	7.4	2.5	-0.5	1.9	5.7	8.1	4.8	1.1	-1.1	3.3	3.4

Source: IMF, *Balance of Payments Statistics Yearbook*, Part 2: World and Regional Tables, Tables B-24 through B-33 (various issues).

differences in rates of return. The observed pecking order of bonds, then bank loans and direct investment is consistent with capital flows between countries populated by "good" firms, efficient banks, and relatively small distortions due to asymmetric information.

b.   Developing Countries

We will focus on net inflows because the volume of capital outflow from developing countries is relatively small.[17] Table 4 illustrates net long-term flows to developing countries over the 1990–1999 period. Capital inflows reached their peak in 1997 at $338 billion. In terms of nominal flows, direct investment steadily increased throughout the sample up to 1997, while the other types of capital flow exhibited much more volatility. Equity and bond inflows reached their peak in the middle of the sample and then declined in the late 1990s. This reversal of portfolio capital flows to developing countries has received a great deal of attention in the literature on emerging markets and capital flows.[18]

The composition of capital flows is shown in Table 5. Two clear shifts in the composition of capital flows stand out from the figures. The first shift is the steady decline of official development assistance as a share of total flow throughout the 1990s, mirrored by the increased share of direct investment. The second shift is the ballooning of portfolio flows in the 1993–1996 period and their subsequent contraction. The pecking order of international finance that seems to emerge from the data is the dominance of FDI flows, followed by bonds, equity, and bank loans. The predominance of FDI flows to developing countries is in part due to the presence of China in the sample, which is a major recipient of direct investment and has restrictions on other types of capital inflows. The final two columns of Table 5 report the decomposition of capital flows to developing countries, excluding China. FDI still leads the pecking order (though with a smaller share) and bank loans remain at the bottom, but the order of equity and bond flows are now reversed.

---

[17] Similar figures on net capital inflows to emerging markets are reported in IFC, *Emerging Stock Markets Factbook* (various issues). The volume of capital flows to the full set of developing countries reported in Table 4 and the volume of flows to the countries included in the set of "emerging markets" is remarkably similar, since very little capital flows to the poorest of developing countries.

[18] To mention just a few of the relevant papers on the reversal of portfolio flows to developing countries, see Calvo (1998), Calvo and Reinhart (1999), and Mendoza (2000).

Table 4: Net Long-Term Flows to Developing Countries (billions of dollars)

	1990	1991	1992	1993	1994	1995	1996	1997	1998	1999
Direct investment	24.1	35.3	47.5	66.0	88.8	105.0	130.8	170.3	170.9	192.0
Portfolio investment	4.0	18.5	25.2	87.6	73.4	66.9	111.6	79.1	53.5	52.6
Equities	2.8	7.6	14.1	51.0	35.2	36.1	49.2	30.2	15.6	27.6
Bonds & money market	1.2	10.9	11.1	36.6	38.2	30.8	62.4	48.9	37.9	25.0
Of which public borrowing:		7.9	2.5	17.5	15.6	12.8	27.9	19.2	27.7	
Bank lending and other debt	14.5	7.8	27.1	12.2	12.3	31.4	39.7	54.6	41.5	-5.9
Of which public borrowing:		0.6	13.7	8.7	7.5	9.8	11.6	14.7	21.9	
Official flows	55.9	62.3	54.0	53.4	45.9	53.9	31.0	39.9	50.6	52.0
Of which public borrowing:		27.2	23.5	25.0	13.1	21.1	3.3	13.3	24.9	
Total	98.5	123.9	153.8	219.2	220.4	257.2	313.1	343.9	316.5	290.7

Source: World Bank, *Global Development Finance* (various issues).

Table 5: Developing Countries: Shares of Total Net Inflows

	1990	1991	1992	1993	1994	1995	1996	1997	1998	1999	Average 1990-93	Average 1994-99	excl. China 1990-93	excl. China 1994-98
Direct investment	24.5	28.5	30.9	30.1	40.3	40.8	41.8	49.5	54.0	66.0	28.5	48.7	24.4	38.1
Portfolio investment	4.1	14.9	16.4	40.0	33.3	26.0	35.6	23.0	16.9	18.1	18.8	25.5	20.8	30.3
Equities	2.8	6.1	9.2	23.3	16.0	14.0	15.7	8.8	4.9	9.5	10.4	11.5	11.3	12.8
Bonds	1.2	8.8	7.2	16.7	17.3	12.0	19.9	14.2	12.0	8.6	8.5	14.0	9.6	17.5
Bank lending and other debt	14.7	6.3	17.6	5.6	5.6	12.2	12.7	15.9	13.1	-2.0	11.1	9.6	10.7	14.3
Official flows	56.8	50.3	35.1	24.4	20.8	21.0	9.9	11.6	16.0	17.9	41.6	16.2	44.1	17.3
	1990	1991	1992	1993	1994	1995	1996	1997	1998	1999	1990-92	1993-98	1990-93	1994-98
Shares of private inflows														
Direct investment	56.6	57.3	47.6	39.8	50.9	51.6	46.4	56.0	64.3	80.4	53.8	51.5	42.8	46.7
Portfolio investment	9.4	30.0	25.3	52.8	42.1	32.9	39.6	26.0	20.1	22.0	21.6	35.6	30.1	26.9
Equities	6.6	12.3	14.1	30.8	20.2	17.8	17.4	9.9	5.9	11.6	11.0	17.0	20.6	15.4
Bonds	2.8	17.7	11.1	22.1	21.9	15.2	22.1	16.1	14.3	10.5	10.5	18.6	9.4	11.5
Bank lending and other debt	34.0	12.7	27.2	7.4	7.0	15.4	14.1	18.0	15.6	-2.5	24.6	12.9	10.3	11.2

Source: World Bank, Global Development Finance (various issues).

Table 6 contrasts the structure of international capital flows in developing and industrialized countries. Given the changes in the composition of capital flows in developing countries over time, the sample is split at end-1992. It is clear that developing countries rely far less heavily on bond flows, consistent with the interpretation that bonds are more attractive to low-risk firms and that investments in developing countries entail greater risk and greater information asymmetries. The data suggest that developing countries rely somewhat more heavily on equity finance, also consistent with the view of greater risk in developing countries. The starkest difference, however, is the heavy reliance on direct investment in developing countries. In the context of our model, direct investment can be interpreted in two ways. If the motivation for cross-border diversification is believed to drive FDI, then direct investment can be thought of as a form of equity investment. On the other hand, if FDI flow is motivated by corporate control, then direct investment can be interpreted as a form of internal finance.

Either interpretation of FDI, as a means of equity diversification or as a means of corporate control, would suggest that FDI flows would be more important in developing countries than in industrialized countries. The steady increase in FDI flows throughout the sample and the mid-sample reversal of portfolio flows, however, is more difficult to explain.

The last two columns of Table 6 list the results from the numerical model for the "benchmark" and "developing" countries (columns 1 and 4 of Table 1). Consistent with the data, the model generates a larger share of equity and a smaller share of bank finance in developing countries relative to the benchmark, though the differences are more pronounced in the data than in the model. The largest difference between the model and the data is that the model generates too large a role for bond finance and predicts that the share of bond finance will be larger in developing countries. An adjustment of parameter values could reduce the relative importance of bond finance. However, the large share of bond finance predicted in developing countries relative to the benchmark is an unavoidable consequence of the interaction of asymmetric information and increased risk.

4.   Other Important Factors for Explaining the Structure of International Capital Flows

The model of the pecking order based on asymmetric information between borrowers and lenders and costly intermediation captured some of the cross-sectional differences in external finance in industrialized and developing countries. However, the model has little to say about short-run fluctuations in capital

Table 6: Structure of Capital Inflows in Industrialized and Developing Countries

	1988–1992[a]		1993–1998		Simulations	
	Industrialized countries[b]	Developing countries[c]	Industrialized countries[b]	Developing countries[c]	Benchmark	Developing country
A. Percent of total flow						
Direct investment	18.0	53.8	19.2	51.5		
Portfolio investment	44.7	21.6	53.6	35.6		
Equity	8.8	11.0	15.2	17.0		
Bonds	36.0	10.5	38.4	18.6		
Loans & other debt	37.2	24.6	27.2	12.9		
B. Percent of total flow excl. FDI						
Equity	10.7	23.8	18.8	35.1	14.8	19.9
Bonds	43.9	22.7	47.5	38.4	55.6	58.8
Loans	45.4	53.2	33.7	26.6	29.6	21.3

[a] Averages for developing countries were calculated over 1990–1993 and 1994–1998. — [b] Percent of net inflows excluding official flows. (Developing country figures exclude China.) — [c] Percent of total inflows.

Source: Authors' calculations based on data reported in Tables 1–5.

flows unless investor expectations or the cost of financial intermediation are assumed to be extremely volatile. The model also produces a stable long-run pattern of capital flows that misses the recent reversal of portfolio flows to developing countries. Finally, our model takes the supply of savings as given and focuses on the allocation of saving across projects. Much of the literature on international capital flows suggests that the volatility in international capital flows may be due to fluctuations in the supply of, rather than the demand for, capital. In this section, we briefly discuss some features of international capital markets that are missing from our model.

Moral Hazard and Investment

In the simple model in Section 2, we assumed that all funds that were borrowed were ultimately invested in the project. If instead the amount of actual investment in the project is left up to the entrepreneur and if creditors are unable to observe the level of investment, a moral hazard problem would arise. In this situation, the optimal capital structure would be tilted toward debt and away from equity in order to shift the risk of a low outcome onto the borrower and improve the incentive for investment. However, Rogoff (1999) argues that even if one believes that moral hazard is an important problem in international lending, global financial markets already disproportionately favor debt over equity.[19] Indeed, in his view, a common feature of the recent proposals to reform the global financial architecture is that they fail to address current biases in the system toward debt at the expense of equity and FDI.

Banking

Our model is extremely naive in that it emphasized the role of banks in monitoring loans and neglected potential issues related to bank runs and self-fulfilling crises. Chang and Velasco (1999a) argue that illiquidity of banks due to a mismatch between short-term obligations and liquid assets can result in a collapse of a financial system. One remedy to prevent bank runs is to provide deposit protection. While protection of deposits prevents bank runs, it further exacerbates the potential for moral hazard. As a consequence, the increased guarantees provided by the government may encourage excessive lending by banks.

---

[19]   Rogoff cites four sources of bias toward debt: taxpayer subsidization of bank intermediation, greater protection of debt contracts in international courts, the underdevelopment of equity markets, and the effective bailout of creditors in industrialized countries by international institutions.

Rodrik and Velasco (2000) argue that maturity mismatch could have played a major role in the Asian crisis. They show that countries with short-term liabilities to foreign banks that exceeded reserves were three times more likely to experience a sudden reversal in capital flows. In their model, runs occur when investors take on large amounts of short-term debt. As in Chang and Velasco, moral hazard induces investors to take on excessive short-term debt when they expect a bailout. Clearly, a more realistic model of banks on the deposit as well as the asset side would alter the implications for the optimal pecking order of international capital flows.

Fixed Exchange Rates

Our model abstracted completely from foreign exchange risk. If debt contracts are written in foreign currency and borrowers do not appropriately price the exchange rate risk, debt contracts may be riskier ex post than they appeared ex ante. Investors may prefer a more liquid form of investment in order to protect themselves from exchange rate risk. If investors are unable to hedge exchange rate risk, this should tilt borrowing toward equity and away from debt.

Government Borrowing/Sovereign Debt

In our model, all financial intermediation took place between private individuals in a competitive financial market. In fact, a significant proportion of capital inflows to both emerging and industrialized countries represented borrowing by governments. According to the World Bank, public debt accounted for over 80 percent of the stock of long-term debt accumulated by developing countries over the past decade. The presence of the government in capital markets could change the equilibrium in a number of respects. First, governments may exercise some market power and they are likely to internalize the impact of a marginal decision to borrow on the risk of the aggregate stock of outstanding debt. Second, creditors may be unable to seize government assets in the event of default. Third, governments have the option of nationalizing assets and unilaterally imposing capital controls. All of these effects would have an effect on the optimal capital structure.

Herding and Contagion

Calvo and Mendoza (2000) and Chari and Kehoe (1997) argue that problems of imperfect information and creditor panic are not limited to banks but may also occur in other financial markets. Such models imply that in the event of a panic, a temporary suspension of trading could prevent the dumping of international stocks based on incorrect information.

FDI as the "Good" Type of Capital Flow

The theory based on asymmetric information and costly monitoring suggests that if FDI involves information transfer it will dominate the pecking order. Indeed, the data suggest that there was a shift toward FDI in late 1990s, possibly induced by the increased riskiness of investment in emerging markets. FDI exhibits less volatility than other forms of capital flows and appears to be more responsive to "fundamentals" and less prone to contagion and herding (World Bank, *Global Development Finance* 1999). Thus, many of the suggested reforms of the global financial system have included mechanisms that would favor FDI at the expense of other channels of cross-border capital flows.

Along with the benefits of increased FDI, it is important to also keep in mind some potential drawbacks. First, FDI may lead to increased specialization. In the absence of other mechanisms for risk sharing, FDI could in principle be welfare-reducing (Hull and Tesar 2000). Second, it is difficult to monitor financial transactions within the firm, so a country may experience capital outflows through derivatives or other off-the-balance-sheet transfers that are not reflected in official figures. Third, FDI may be skewed toward particular industries or toward large firms (Blomstrøm and Lipsey 1991). Finally, to the extent that FDI takes place through investment by multinational firms, FDI may make it increasingly difficult for small countries to monitor and regulate firms operating in their countries.

The presence of asymmetric information could also distort the level of investment. Razin et al. (1999) argue that because foreign direct investors have inside information about the firm, they will overcharge uninformed domestic savers for stock in the firm. In anticipation of this excess profit, multinationals will overinvest in the local economy.

5.   Conclusions

The purpose of this paper was to formalize a simple model of borrowing and lending in an open economy when borrowers have access to a menu of options for finance. The model yielded a "pecking order" of finance depending on the distribution of risk across firms and the cost of financial intermediation. In financial autarky, countries with lower costs of financial intermediation and higher expected output will rely more heavily on bond and bank finance. When countries with similar costs of financial intermediation and a similar distribution of firms become integrated into the global capital market, the structure of capital inflows will mirror the structure of capital outflow. When small countries with relatively risky populations of firms integrate into the global capital market, capital inflows will be dominated by FDI and equity inflows. We find that these predictions are broadly consistent with data on international capital flows in industrialized and developing countries.

Our model emphasized the demand side for corporate finance. In future work we plan to incorporate important features of the supply side of finance. We highlighted several aspects of international capital flows that are not accounted for in this demand-side model. Issues such as herding, weakness in the banking sector, and sovereign debt have implications for the supply of international funds. We hope that an accurate model of the supply side will help further explain the composition of capital flows in both emerging and industrialized countries.

Bibliography

Blomstrøm, M., and R.E. Lipscy (1991). Firm Size and Foreign Operations of Multinationals. *Scandinavian Journal of Economics* 93(1): 101–107.

Bolton, P., and X. Freixas (2000). Equity, Bonds, and Bank Debt: Capital Structure and Financial Market Equilibrium under Asymmetric Information. *Journal of Political Economy* 108(2): 324–351.

Calvo, G.A. (1998). Capital Flows and Capital-Market Crises: The Simple Economics of Sudden Stops. *Journal of Applied Economics* 1(1): 35–54.

Calvo, G.A., and E. Mendoza (2000). Rational Contagion and the Globalization of Securities Markets. *Journal of International Economics* 51(1): 79–113.

Calvo, G.A., and C. Reinhart (1999). When Capital Inflows Come to a Sudden Stop: Consequences and Policy Options. In P. Kenen, M. Mussa, and A. Swoboda (eds.), *Key Issues in Reform of the International Monetary and Financial System.* Washington, D.C.: International Monetary Fund.

Chang, R., and A. Velasco (1999a). Financial Crises in Emerging Markets: A Canonical Model. *Federal Reserve Bank of Atlanta Quarterly Bulletin* 84(2): 4–17.

— (1999b). Liquidity Crises in Emerging Markets: Theory and Policy. NBER Working Paper 7272, Cambridge, Mass.

Chari, V.V., and P. Kehoe (1997). Hot Money. NBER Working Paper 6007, Cambridge, Mass.

Collins, S., and B. Bosworth (1999). Capital Flows to Developing Economies: Implications for Saving and Investment. *Brookings Papers on Economic Activity* (1): 143–169.

Corsepius, U., P. Nunnenkamp, and R. Schweickert (1989). *Debt versus Equity Finance in Developing Countries: An Empirical Analysis of the Agent-Principal Model of International Capital Transfer.* Kieler Studien 229. Tübingen: Mohr Siebeck.

Edwards, S. (1999). How Effective Are Capital Controls? NBER Working Paper 7413, Cambridge, Mass.

Hull, L., and L.L. Tesar (2000). Risk, Specialization and the Composition of International Capital Flows. Working Paper, University of Michigan.

IFC (International Finance Corporation) (various issues). *Emerging Stock Markets Factbook.* Capital Markets Department. Washington, D.C.

IMF (International Monetary Fund) (various issues). *Balance of Payments Statistics Yearbook.* Washington, D.C.

— (various issues). *International Financial Statistics.* Washington, D.C.

Institute for International Finance (2000). Recovery Anticipated in Flows of Private Capital to Emerging Markets this Year. Press Release dated January 24, 2000, Washington, D.C.

Lane, P.R., and G.M. Milesi-Ferretti (2001). External Capital Structure: Theory and Evidence. This volume.

Mendoza, E. (2000). Credit, Prices, and Crashes in Emerging Economies: Sudden Stops Economics in an Equilibrium Framework. Working Paper, Duke University.

Modigliani, F., and M.H. Miller (1958). The Cost of Capital, Corporation Finance and the Theory of Investment. *American Economic Review* 48(3): 261–297.

Montiel, P., and C. Reinhart (1999). Do Capital Controls and Macroeconomic Policies Influence the Volume and Composition of Capital Flows? Evidence from the 1990s. *Journal of International Money and Finance* 18(4): 619–635.

Myers, S.C., and N.S. Majluf (1984). Corporate Financing and Investment Decisions When Firms Have Information That Investors Do Not Have. *Journal of Financial Economics* 13(2): 187–221.

Razin, A., E. Sadka, and C.-W. Yuen (1998). Channeling Domestic Savings into Productive Investment Under Asymmetric Information: The Essential Role of Foreign Direct Investment. CEPR Discussion Paper 1837, London.

— (1999). Excessive FDI Flows under Asymmetric Information. NBER Working Paper 7400, Cambridge, Mass.

Rogoff, K. (1999). International Institutions for Reducing Global Financial Instability. NBER Working Paper 7256, Cambridge, Mass.

Rodrik, D. (1998). Who Needs Capital-Account Convertibility. In S. Fischer et al., *Should the IMF Pursue Capital-Account Convertibility?* Essays in International Finance 207. Princeton University.

Rodrik, D., and A. Velasco (2000). Short-Term Capital Flows. In B. Pleskovic (ed.), *Annual World Bank Conference on Development Economics 1999.* Washington D.C.: World Bank.

Schnitzer, M. (1995). Debt Versus Foreign Direct Investment: The Impact of Sovereign Risk on the Structure of Capital Flows to Developing Countries. Discussion Paper 484, Sonderforschungsbereich 303, Universität Bonn.

Schuknecht, L. (1999). A Trade Policy Perspective on Capital Controls. *Finance & Development* 36(1): 38–41.

World Bank (various issues). *Global Development Finance: Country Tables.* Washington, D.C.: World Bank.

Comment on Leslie Hull and Linda L. Tesar

(Selected Country)

Helmut Reisen

F21
F31
F32
O16 G19

1.   The Aim of the Paper

Despite the remarkable rise of private cross-border capital flows over the past decade, their composition remains ill-explained. The potential return of studies in that area is considerable. Leslie Hull and Linda Tesar first announce (p. 88) that their "intention is to develop an analytical framework for thinking about the determinants of the composition of international capital flows." Towards the end of the paper we get a "model of the pecking order [of corporate finance] based on asymmetric information between borrowers and lenders and costly intermediation [that] captured some of the cross-sectional differences in external finance in industrialized and developing countries" (p. 103). A case of diminished expectations as the model fails to explain the volume of capital flows and as it follows earlier attempts at the IMF (not quoted), notably by Razin et al. (1995) and by Chen and Khan (1997), to explain the mix of flows on the basis of information asymmetries and financial intermediation costs.

The model advanced by Hull and Tesar predicts that firms with good credit risks will prefer to raise capital through the bond market, that medium-risk firms unable to tap the bond market will rely on bank loans and/or equity, and that firms with poor credit ratings will rely on equity finance. The basic assumptions that underlie these predictions are that bondholders have priority claims over shareholders, that equity finance includes a risk premium to account for "lemon" firms (which are assumed to be undistinguishable to prospective investors), and that bank finance comes with the flexibility of restructuring and the possibility of information-sharing between the firm and the bank, but entails a monitoring cost reflected by the intermediation spread. Translated for the purpose of cross-border trade, countries populated with high-growth firms and characterized by a relatively high degree of corporate transparency will show a pecking order of bonds, then bank loans, and finally equity investment in their capital account. This pattern should hold for most OECD countries. For developing countries, by contrast, we should observe a higher degree of FDI finance, which minimizes information risks, relative to other capital flows, in capital inflows rather than other forms of private finance.

2. The Empirical Content

Any model providing a positive theory on the mix of private flows will have a difficult job surviving empirical tests of falsification. Bosworth and Collins (1999), for example, examining the pattern of private capital flows for 58 developing countries over the period 1978–1995, find that foreign direct investment, portfolio flows, and bank loans are not significantly correlated with one another over time or across countries. Some salient facts in a nutshell:

— The pattern of inflows differs markedly by region. While bank loans were the dominant type of flows to Latin America prior to 1982, capital inflows have been concentrated in FDI and portfolio capital since then. East Asia first relied mostly on FDI and only in the 1990s experienced strong growth in portfolio flows and bank loans, until the region succumbed to creditor panic in 1997–1998.
— There is little complementarity between the broad types of inflows in the numbers. China, the largest developing-country recipient of FDI in the 1990s, obtained little portfolio capital or lending (like Singapore earlier on). Korea financed its development mostly through bank loans, while discouraging FDI and portfolio equity inflows. Brazil, the largest recipient of portfolio capital among developing countries, reduced its reliance on loans.

Figure 1 contains good and bad news on the empirical content of the Hull–Tesar model. It provides a snapshot for 1996, a noncrisis year for which the number of observations on sovereign country ratings are maximized, to show the correlations between ratings and the share of different types of flows in total capital inflows for 21 developing countries. The good news is that the share of both foreign direct and portfolio equity flows moves higher when ratings, a proxy for default risk, deteriorate. By contrast, portfolio bond flows and bank lending show more prominence in the recipient countries' capital accounts when ratings are better. The bad news, however, is the low explanatory power of ratings for the mix of inflows, as the R^2 does not go beyond 0.03 in the four panels displayed in Figure 1. Hence, we seem to have a problem of missing variables.

3. Information Asymmetries and Corruption

Exploring ways to help improve the explanatory power of the Hull–Tesar model, I confine myself to parameters that are closely related. Razin et al. (1995) use the cost-of-financing argument to explain different forms of capital flows, finding

Figure 1: The Mix of Capital Inflows and Country Ratings, 21 Observations for 1996

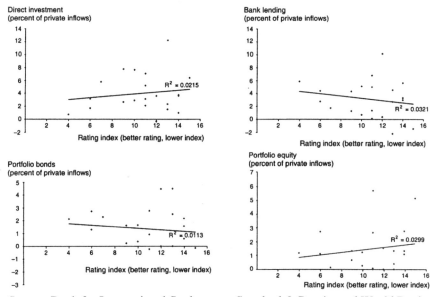

Source: Bank for International Settlements, Standard & Poor's, and World Bank.

"green field" FDI to be least costly, followed by debt flows and then by portfolio equity flows. FDI is less costly as the participation in the management reduces the asymmetric information problem. Chen and Khan (1997) derive their results from the inefficiency of the domestic financial market in the recipient country, which is modeled as a result of asymmetric information between outside investors who rely on information in the domestic financial market and insiders in firms. Their analysis allows predictions to be made on the mix of flows based on a host country's growth potential and financial market development. Countries where the growth potential dominates the degree of financial market development will receive more FDI than portfolio equity flows; countries with suitable parameter values for both growth potential and financial markets will see relatively more equity inflows. The Chen–Khan model allows for sudden reversals of capital flows for economies experiencing changes in the perceived growth potential or financial market integrity, or both.

However, theory and evidence presented in a recent paper by Wei (2000) for the OECD Development Centre seems to contradict the predictions of the information-asymmetry approach presented above, including those by Hull and Tesar, if you accept that local information and corruption problems are correlated. Wei presents strong empirical evidence that countries with high corruption indices have a relatively low share of FDI in their capital imports, while bank and portfolio flows are unaffected by corruption levels in the host country.

Table 1 is based on a generalized gravity model specification

$$\log (\mathrm{loan}_{jk}/\mathrm{FDI}_{jk}) = \text{source country} + \beta \text{ corruption }_k + x_{jk} \zeta + e_{jk},$$
$$\text{fixed effects}$$

where β and ζ are scalar and vector parameters, respectively, and x_{jk} is a vector of determinants of bilateral FDI other than host country corruption. Wei also finds for U.S. data that host country corruption measures affect the ratio of portfolio flows to FDI positively as well.

Table 1: Corruption and the Mix of Capital Inflows

	Log (loan/FDI)	
	Fixed effects	Random effects
WDR corruption measure	0.793**	1.228**
	(0.328)	(0.615)
Log GDP	−0.333**	−0.476**
	(0.114)	(0.212)
Linguistic tie	−0.705**	−0.504**
	(0.335)	(0.291)
R²	0.37	0.39
No. of observations	197	197
Breusch/Pagan test Prob $> \chi^2$		0.00
Hausman test Prob $> \chi^2$		0.95

Note: Log GDP per capita and log distance between economies were further determinants, which are not reported here, as they did not enter significantly. Here the *World Development Report 1997* corruption measure is displayed; measures from the *Transparency International* and from the *Global Competitiveness Report 1997* performed significantly as well.

Source: Wei (2000).

These findings seem important to me as they may indicate that the effects of information asymmetries between investors and recipients have often not been appropriately specified. First, how do international direct investors obtain the presumed informational advantage over bank lenders and portfolio investors? Largely, by greater direct exposure to the host country, including by sending managers from headquarters to the country. International direct investors are thus more likely to have repeated interactions with local officials (for permits, taxes, health inspection, and so forth) than foreign banks or portfolio investors, raising the need to pay bribes and to deal with extortion by local bureaucrats. Second, direct investment involves greater sunk costs than bank loans or portfolio investment. This puts direct investors in a weaker bargaining position than investors in more liquid assets. This ex post disadvantage of FDI makes international direct investors more cautious than international portfolio investors ex ante to raise their claims on a corrupt host country.

4.   Policies Matter

Models of the structure of capital flows stripped of policy response functions to capital inflows fail to hold much empirical content. The monetary-fiscal policy mix, including the exchange rate regime, of the host country can be expected to significantly determine the mix of inflows; tax-subsidy regimes may play an important role as well. In many instances, domestic policies may have been designed deliberately to influence the structure of capital inflows, for example, through (dis)incentives and restrictions.

Montiel and Reinhart (1999) present evidence that capital controls influence the composition of flows, not their volume, towards FDI, while sterilized intervention influences volume and composition, skewing flows to short maturities. They estimate a set of fixed-effect panel regressions explaining the volume and composition of inflows as a function of sterilization, the severity of capital account restrictions, international interest rates, and a proxy for capital market depth. The estimate covers 105 annual observations for 15 emerging economies over the years 1990–1996. Montiel and Reinhart also find, not surprisingly, that the depth of local stock markets is positively correlated with the share of portfolio equity inflows, but insignificant for other types of flows.

Excessive reliance on short-term borrowing, the single most important trigger of recent currency crises in emerging markets, can be discouraged by flexible exchange rates. By contrast, exchange rate pegs, in combination with high interest rates, typical in developing countries for structural reasons, tend to reinforce bank lending and spending booms (Reisen 1998). They constitute an incentive for leveraged investors to exploit interest differentials as well as for offshore borrowing by creditworthy banks and nonbanks to tap seemingly cheap sources of finance. Central bank intervention on the foreign exchange market to peg the currency in the face of net inflows, unless sterilized fully, is intermediated into the domestic banking system. The exchange rate peg provides the incentive to allocate those funds disregarding currency and maturity risks, as these are being implicitly transferred to the central bank. Keeping nominal exchange rates flexible, even introducing "noise" through central bank intervention when it is seen to be on a too-stable, appreciating trend during inflow periods, improves the mix of inflows towards longer maturities and encourages banks and firms to hedge their foreign-currency exposures.

Most countries seem well advised to encourage FDI and portfolio equity investment and to avoid any implicit subsidies for debt-creating flows. Reisen and Soto (2001) measure the independent growth impact of the various broad categories of private inflows, providing a panel data analysis covering 44 developing countries over the period 1986–1997. Their findings suggest that both FDI and portfolio equity investment exert a significant independent growth effect (after correcting for other growth determinants). Bond flows do not enter significantly in the growth regressions. Foreign bank lending is shown to contribute to growth only if the local banking system is well capitalized, otherwise its independent growth effect is shown to be negative.

Bibliography

Bosworth, B., and S. Collins (1999). Capital Flows to Developing Economies: Implications for Savings and Investment. *Brookings Papers on Economic Activity* (1): 143–180.

Chen, Z., and M. Khan (1997). Patterns of Capital Flows to Emerging Markets: A Theoretical Perspective. IMF Working Paper WP 97/13, Washington, D.C.

Montiel, P., and C. Reinhart (1999). Do Capital Controls and Macroeconomic Policies Influence the Volume and Composition of Capital Flows? Evidence from the 1990s. *Journal of International Money and Finance* 18(4): 619–635.

Razin, A., E. Sadka, and C.-W. Yuen (1995). A Pecking Order Theory of Capital Inflows and International Tax Principles. IMF Working Paper WP/96/26, Washington, D.C.

Reisen, H. (1998). Domestic Causes of Currency Crises: Policy Lessons for Crisis Avoidance. OECD Development Centre Technical Paper 136, Paris.

Reisen, H., and M. Soto (2001). Which Types of Capital Inflows Foster Developing-Country Growth? *International Finance* 4(1).

Wei, S.-J. (2000). Corruption, Composition of Capital Flows and Currency Crises. OECD Development Centre Technical Paper 165, Paris.

Gunter Dufey

G21   G28
F32   F23

The Blurring Borders of Banking

/ US /

1.   Introduction: Context and Significance

Capital flows do not take place in a vacuum. They are channeled through institutions that constitute an important part of the world's financial landscape. This institutional framework has a significant impact on the magnitude, the direction, and the volatility of capital flows.

Issues related to the structure and the dynamics of the institutional framework of the global financial system attracted renewed attention in the 1990s. The world has learned that "financial system architecture" (Boot and Thakor 1997) can and does affect economic performance in significant ways in terms of both quantity—in the sense of attaining growth potential—as well as the quality—in the sense of macroeconomic stability. While a well-functioning financial system can yield growth rates on the basis of very modest savings rates by minimizing the misallocation of resources, as we saw in the United States during the 1990s, a nonfunctional system causes persistent value destruction, as shown by the Russian economy during the last decade. The Asian crisis, finally, has taught everyone a lesson about the role of the financial system in ascertaining macroeconomic stability, or rather the lack thereof.

Traditionally, banks have constituted the core of the financial system. In some countries they have played a dominant role (e.g., Japan); in others, they have had to share the delivery of financial services[1] with many nonbank financial intermediaries as well as independent capital market institutions, depending on the country's historical/political evolution and the resulting regulatory framework.

---

[1]   For purposes of this paper the term "banking services" is used in the traditional, narrow context of commercial banking services, e.g., the execution of payments, time deposit accounts for savers, various overdraft facilities, and loans. In contrast, the term "financial services" takes a comprehensive, broad view of financial products, including insurance, leasing, securities broking, and other portfolio investment products and their distribution.

To illustrate this point, in the United States, for political reasons deeply imbedded in the history of that country, the power of banks has been deliberately curtailed, through regulatory intervention, as exemplified by the McFadden and the Glass–Steagall Acts. In line with this tradition, the United States also defined banks in a very narrow sense, i.e., as "deposit taking institutions."

In contrast, Europe and Japan chose a different path. The European Union, for example, uses a much broader definition: banks comprise all "monetary financial institutions" whose business is to receive deposits and close substitutes for deposits (for example, through the issuance of debt securities) and, for their own account, grant credit (including investing in securities). The consequences in terms of the shape of the institutional framework are substantial, as comparative analyses show (Dufey 1998b and sources cited therein). None of these observations is new and there exists an extensive literature. What the analysis of the legal framework shows, however, is that the institutions which make up a financial system have been shaped over time by a complex combination of the interaction between (a) microeconomics, (b) systemic "accidents," and (c) regulatory response. In other words, extant financial systems are path dependent.

Technological change over the last 20 years or so has accelerated the process. The purpose of this paper is to show the driving forces of this change and ultimately explore the results in terms of the borders of banks in particular and the financial services industry in general. Such an examination begins logically with a look at the traditional core of the community of financial service providers, i.e., institutions called "banks," before proceeding with an analysis of the blurring of the borders of such enterprises.

a.   The Theory of Banking and the Uniqueness of Banks

Clearly, rapid technological progress has permitted a host of new entrants into the market for financial services, which were traditionally the domain of banks. Of course, this phenomenon has given rise to the question whether there are any areas of banking that represent a field of unique competency of such institutions. Banks and banking have been the subject of intense academic analysis for a very long time. However, under the seemingly sudden onslaught of new competitors and the arrival of new forms of delivering financial services, the issue of what is the essence of banking has gained a new urgency.

For a long time banking theory has been dominated by the notion that these institutions possess information-based advantages. Monitoring the transaction accounts of their customers, banks obtained unique insights as a matter of course. Such information, in turn, was believed to allow them to arrive at superior judgements about the financial services needs and the creditworthiness of their customers. With the spread of information technology, data mining, and similar techniques accessible to many other institutions, this traditionally held advantage of banks has been weakened considerably. Indeed the question has been raised as to what extent banks are still necessary at all (see Llewellyn 2000 and sources cited therein). This challenge consequently gave rise to the search for a revisionist view of the essential advantage of banks.

The evolving theory of banking of the 1990s arrives indeed at a consensus about banking: *yes*, there is something unique about banks. It is they who provide "liquidity services," i.e., options on readily available purchasing power, to both businesses and consumers.

While previous theories of banking have focused on either the deposit taking or, alternatively the lending function of banks (for a concise presentation of the literature, see Bhattacharya and Thakor 1993), the emerging view of the 1990s recognizes explicitly the *relatedness* of these functions. On both sides of the balance sheet, such intermediaries, i.e., banks, capture unique advantages because of their ability to reduce uncertainty by pooling, by capturing scale economies, and, last but not least, by exercising greater market power than that exerted by individuals. In a recent paper (Kashyap et al. 1999), it is recognized that the importance of banks is based not so much on the loans that they disburse but, indeed on their loan *commitments*, i.e., the options on liquidity that they offer their customers. Thus, both on the deposit side and the lending side, banks offer similar services—often to the very same customers: namely the provision of purchasing power on demand in response to unpredictable requirements for funds. Both the deposit business as well as the provision of loan commitments require "reserves" in the form of cash and securities holdings. The synergistic effect between the two activities becomes particularly clear when it is recognized that the timing of liquidity demands of depositors and borrowers are less than perfectly correlated.

The concept of the uniqueness of banks based on the provision of liquidity would be incomplete, however, without an inherent institutional feature of modern financial systems: the efficiency of banks in providing liquidity services is further enhanced by the fact that they have direct access to liquidity reserves at the respective central bank. Thus, they not only have available liquid reserve accounts which are default free, but they tend to have access to overdraft facilities (see Corrigan 1982 and 2000). It is true that such facilities are often available on a discretionary basis only, but almost always at preferential rates. In return, banking firms must accept a special regime of supervision.

This public policy dimension of banking gives the concept of defining the institution "bank" a certain arbitrary aspect: banks are financial intermediaries that perform a unique economic function for other entities in the economy. Yet, whether they can perform this function competitively depends ultimately on a discretionary regulatory decision: *exclusive, direct access to central bank liquidity!*

Recently, the issue how further advances in technology affect banks and central bank powers has come to the fore, particularly with respect to the growing introduction of e-money by nonbank institutions such as web-based auction houses, firms that provide travel and other services electronically, or even other entities who command a degree of operational competency and integrity. The resulting debates have yielded a number of interesting insights that parallel in some respects earlier controversies regarding the role of nonbank financial intermediaries. The outcome of the debates is likewise quite similar: First, e-money is unlikely to replace cash for transactions completely, for a variety of reasons, mainly having to do with the privacy motive. And as long as there is some cash in the system, central banks will be able to influence the "short-term rental fee" for liquidity, i.e., interest rates. Second, indeed if all cash in an economy were to be replaced by e-money, the respective central bank would still be able to influence short-term interest rates by offering to lend or borrow e-money at a rate below or above the prevailing time value of purchasing power.[2] Such activities may affect a central bank's profitability, but not its essential ability to control the supply of liquidity to an economy.

The conclusion from the review of traditional banking can be generalized. The *interaction* between economic performance and regulatory discretion, while not totally unique to banking, is a significant aspect of all the institutions that comprise the financial system as a whole. This becomes clear when one looks at the business of providing financial services in a broader context.

b.   Other Financial Service Providers

Generic FSPs (financial service providers) focus on delivering services which involve originating contracts, gathering, analyzing and disseminating information, managing risks, and in the process establishing and maintaining relationships with customers and other FSP's. Depending on the technology available, which determines product cost, a wide variety of institutions have emerged that meet two conditions: (1) the ability to produce and deliver a financial service (or bundle of services) competitively *and* (2) being licensed to operate in a given market

---

2   For an extensive analysis, see papers by C. Goodhart (2000) and others, published as Conference Proceedings in *International Finance*. See also the similar conclusions of earlier work published in Deutsche Bundesbank, *Monthly Report*, June 1999 and March 1997.

(see Table 1). Thus, it is the *interaction* of economic/managerial capability with the regulatory framework that results in the institutional structure of a financial system.

Table 1: Structure of Financial Institutions

United States	Europe
Commercial banks	Banks[a]
Savings institutions	Banks[a]
Credit unions	Banks[a]
Finance companies	Banks[a]
Securities brokerage	Banks[a]
On-line brokerage	Banks[a]
Investment banks	Banks[a]
Mutual fund companies	Affiliates of banks
Mortgage companies	Mortgage banks
Insurance companies	Insurance companies
[a] Differences due to ownership.	

2.   Dynamic Factors Driving Changes in the Banking Industry[3]

The objective of this section is to show the causal links between the drivers of change in the banking industry. Beginning with a brief review of developments regarding relevant technology, I look at the impact on the internationalization of the financial service business. I will show that this internationalization process had a major impact on the regulatory environment. By allowing financial institutions to choose a regulatory regime for all or parts of their business, internationalization became a driving force with respect to the process of deregulation and liberalization that occurred in world financial markets, albeit at various speeds in different countries.

Regulatory liberalization has many effects on the financial service industry. The essence of these effects is reflected in the pace of financial innovation. In

---

3   Parts of this section are based on Dufey (1998a).

turn, financial innovation has profoundly changed cost structures within and between institutions for creating and delivering financial services. The result is an ongoing reshuffling of the shape and scope of the institutions that produce and deliver such services. It is at this point that the real essence of the blurring borders of banking becomes apparent. Finally, I will emphasize the dynamic nature of this process.

a.   Technology and Its Impact on Financial Services

Of the multiple and complex aspects of technology, two appear of particular importance for a business such as financial services, whose essence consists of contracts and bits of information. One is computational capability that allows the manipulation of numerical data at very high speeds. The other aspect of technology that is relevant here is communications technology. This allows the transmission of large amounts of data over long distances at ever-decreasing costs, making geographical distances less and less relevant. Obviously, the latter phenomenon has given rise to the internationalization, or even globalization, of financial services.

b.   Internationalization/Globalization

With advances in communications technology, it was no surprise to find that financial claims began to cross borders in the post-world-war period, in spite of the fact that during wartime, extensive regulations were imposed on transactions. "Currency inconvertibility" is the term that characterized the period. However, it was recognized early on that economic recovery and growth require a modicum of freedom to conduct financial transactions, and once the regulatory environment for those transactions had been sufficiently liberalized, e.g., convertibility of current account transactions had been introduced, the process could not be curtailed. Once the dikes had been breached, money, like water, created ever widening gaps in the bulwark of capital controls. The advances in information and communications technology allowed financial institutions as well as their customers to look for more hospitable environments.

   A dramatic example is the growth of the so-called offshore markets. The essence of this phenomenon was reflected in the fact that substantial proportions of the total credit intermediation activity for major currencies has shifted onto the books of banks outside the country where the respective currency is means of payment. To illustrate this point, by the 1990s, more than 50 percent of all dollar *time deposits* were found to be on the books of banks *outside* of the United

States. Although many of these institutions are simply branches of U.S. banks, often no more than "mailbox entities" in some convenient offshore jurisdiction, many are foreign-based banks who are now enabled to competitively offer deposits and loans denominated in U.S. dollars. For other convertible currencies the proportions of funds intermediated offshore are somewhat smaller, but still significant.

In practical terms, the advantage of the offshore market means that every transactor, resident or nonresident, has increasingly had the alternative of contracting for a time deposit in a market outside the country in which the respective currencies is means of payment. The same is true on the credit side. Any potential borrower now has a choice; funds can be obtained from a financial institution in the national market of their currency, or a market outside that country in the very same currency. It is not surprising that these developments have increased competition between banks dramatically. Essentially, through the offshore markets, it became possible to separate currency, jurisdiction, and institution. To illustrate this point: anyone with significant liquid funds can obtain a yen (time) deposit in a branch of a Canadian Bank in, say, Singapore. Of course the same holds on the borrowing side. Indeed, this phenomenon can well be considered the true essence of the globalization of financial markets, the disappearance of the links that traditionally define a national financial market: currency, regulatory framework, and the governance of financial institutions.

c.   **Deregulation and Liberalization**

It is a commonplace to state that banking and finance are regulated industries. In virtually all countries, the role of government extends far beyond the traditional scope of economic regulation, i.e., the enforcement of contracts and basic consumer protections. Three reasons are traditionally put forth to justify the need for a more extensive regulation of the financial market place relative to the markets for most other products: (a) banks are institutions that offer demand deposits that constitute the primary means of payments, (b) the safety of the institution must be protected because of the possible secondary effects of the default of financial institutions in the nonfinancial sector of the economy, and (c) to assure an "appropriate" or "just" allocation of credit.

Over time, public debate initiated in academic writings has found most of these arguments wanting. It has been recognized that the quest for regulation in a political economy originates primarily from two sources: (1) Established competitors wish to maintain their market positions, and regulation is definitely an effective way to defend established market shares from the onslaught of new entrants in the industry. (2) The other powerful force for financial market regulation is represented by the political establishment, whose members find the financial system a wonderful opportunity to allocate resources according to political imperatives rather than the rules of economic efficiency. Obviously, it is at this point where internationalization comes into play. Financial institutions and their customers have the opportunity to escape such constraints by moving their activities to alternative jurisdictions.

The resulting arbitrage compels a rethinking of the needs for regulation. Put differently, the question surfaces in terms of the *economic benefits* of regulation. The body politic begins, albeit slowly, to distinguish between regulations that create value by making the system more efficient and those that only serve special interests but detract from the benefits accruing to the economy at large. Regulatory economic value is created by rendering the system more efficient by making it a safer and more transparent place to do business. Conflicts between individual institutions and society at large must be reconciled at this juncture.

It is not surprising then that the globalization of markets, as it is often referred to, has put tremendous pressure on different national regulatory systems. It has been the ultimate source for further liberalization, a phenomenon whose success is difficult to understand in the face of resistance not only from established competitors, but also the bureaucracies and members of the political class who lose power and influence in an important dimension of the political economy, namely the allocation of financial resources. In such an environment, regulators are forced to focus on the true need for regulation.

Globalization of competition in financial services has in turn brought to the fore the need to harmonize and coordinate regulatory systems across borders. The confrontation of various national regulatory regimes in combination with the pressure to harmonize has forced regulators to find common ground by restricting themselves to focus on those regulations which are really essential, i.e., those that indeed create value by enhancing the safety and efficiency of the system. In contrast, regulations that have negative economic value by detracting from the efficiency and safety of the system tend to be eliminated. A perfect example is provided by rules stipulating minimum reserve requirements that exceed the working balances of banks with a central bank.

A review of liberalization efforts in financial markets during the last thirty years and more recently in many emerging markets is illustrative of these dynamic effects. Clearly, the trend toward deregulation has provided the scope for banks and other financial institutions to enlarge the offerings of their products and to offer them in different forms. Recent changes in the United States illustrate this. New legislation in the form of the Gramm–Leach–Bliley Act (see Appendix) was put into effect in 1999, considerably expanding the permissible scope of operations and services for U.S. banks.

With respect to the liberalization process in financial markets of emerging markets and transition economies, a troubling observation has vexed policy makers and indeed a wide range of financial market participants: it appears that the process of (external) financial market liberalization advocated strongly by most industrialized countries and the international institutions such as the IMF and the World Bank has been accompanied by subsequent financial crises. This empirically observable pattern has been so pervasive that a causal relationship has been suggested.

Closer analysis of the circumstances of such crises has provided evidence that early analyses of the liberalization-financial crises linkage tended to overlook a crucial aspect of this process: in almost all countries afflicted by this phenomenon, it turned out that *external* liberalization was not matched by *internal* financial market liberalization. The credit allocating banking institutions continued to be influenced disproportionately by government and/or related nonbank entities. The absence of proper incentives for prudent lending, combined with unrestrained access to funds from abroad, and a regulatory system that made for moral hazard by providing implicit or even explicit government guarantees could not help but create conditions that were prone to systemic bank failure.

d.   Financial Innovation

Historically, it is not surprising that the wave of financial innovation that has characterized financial markets during the past two decades arrived shortly after markets of industrialized countries had become internationally integrated and had gone through a thorough process of de jure and de facto liberalization. Banks and other financial institutions began to explore new products and to experiment with new delivery systems in response to forces on the demand side as well as the supply side. On the former, market turbulence that accompanied deregulation stimulated demand for hedging products. At the same time, the supply of such products was facilitated by low-cost computing power that quickly became widespread. Further, and probably for first time in history, the contributions from academia had a significant impact in providing formal models of new instruments allowing for hedging and therefore relatively precise pricing of such instruments.[4]

---

[4]   The most prominent example is options. While records of the use of such financial contracts go back to the 15th century, where options on commodities were traded in Osaka and later Amsterdam, only the relatively recent work of Black, Scholes, and others permitted institutions to offer such products as an integral part of their business.

The process of financial innovation which began in the United States in the early 1980s, reaching Europe in the mid-1980s and establishing itself in Japan in the early 1990s, has been the subject of an extensive literature. There is indeed evidence that the design of a financial system has a distinct impact on its ability to innovate (Boot and Thakor 1996). The process of financial innovation per se comprises essentially three distinct aspects: the unbundling and bundling of financial contracts, the "securitization" of illiquid claims, and the development of new channels for distributing financial services via telephone and the internet, the mobile dimension representing the most recent refinement of technological progress.

The unbundling/bundling technique of financial innovation starts with the premise that traditional financial contracts, say a fixed rate loan, really consist of a bundle of different "instruments" which, with the help of some computational capabilities, can be "stripped" into a variety of components: among others there is the *availability guarantee* that has become incorporated in various "commitment" instruments such as revolving underwriting and note issuance "facilities." Closely related are a variety of credit derivatives. Further, it is possible to strip out the commitment of a fixed interest rate, isolate it, and trade it as *interest rate futures*, *FRAs*, and *swaps*. This is not all: when borrowers can get out of fixed rate obligations by law or simply market power, the lender has granted an *interest rate option*. And to the extent that assets whose price correlates with the price of a commodity collateralize the loan in one way or another, we are dealing de facto with a *commodity option*.

The economic advantage of this process is based on the potential of selling the various risks and "burdens" to those in the marketplace who charge least for them, lowering total costs in the process. By the same token, however, it must be recognized that this activity facilitates arbitrage transactions that leave some of the traditional and slower reacting institutions with those elements of the "bundle" where compensation is inadequate relative to the risks taken.

Just as easy as traditional contracts can be unbundled, it is possible to combine various elements, but now into packages that fit more precisely the needs of market participants who will be paying only for what they really want and not for "packages" with elements of little or no interest to them. This of course is the essence of financial engineering that has become the defining trademark of the modern investment banking industry.

Another dimension of financial engineering involves the process of securitization: the "repackaging" of illiquid financial claims as tradeable securities. The classic illustration of this principle is represented by the development of the negotiable CD that substituted for an illiquid time deposit. The economic benefits of securitization are obvious. Economic agents value liquidity because it allows reaction to unexpected needs for funds and it permits them to get out of value-losing positions quickly, before other market participants have heard the "news." Obviously, the latter is mostly an illusion, especially in light of ever more efficient communications—but then people are willing to pay for illusions.

Much has been written and said about the arrival of new delivery systems for financial services (for an excellent survey, see *Online Finance: The Virtual Threat* in *The Economist*, May 20th, 2000). Clearly, by changing the marginal cost for the most mobile customers, these new technologies have an impact, on both the individual firms and the structure of the industry, that goes far beyond their relatively small market penetration in even the most advanced countries. What is important, however, is that these technologies have rendered markets *contestable,* thoroughly changing the competitive dynamics in the market for financial services.

e.   Changing Institutional Structures, "Blurring" Borders

From the forces of change identified above, for observers of the rapidly changing industrial structure of the financial services industry, a confusing pattern of crosscurrents becomes apparent: very few generalizations hold up to scrutiny if one reviews global markets for financial services and their providers (Walter 1998). Indeed, looking at two major dimensions of "diversification" by geographic reach and scope of product offering, a seeming contradictory pattern emerges. While a handful of U.S. investment banks has achieved significant market shares by the second half of the 1990's, hovering around 50 percent in Tokyo, Frankfurt, Paris, and Madrid, not to mention London, their commercial banking brethren rarely attain more than 5 percent market share in terms of local business, which includes the local affiliates of multinational companies headquartered in their respective home countries. Those that hail from other countries do even worse in markets outside heir home countries. Significant exceptions to this picture can be found only in emerging markets, where foreign commercial banks enjoy technological and reputation advantages against local competitors, weakened by eons of abuse by their governments who forced them to channel funds into politically attractive projects.

At the same time, one observes a constant shift in focus: while a number of financial firms attempt to turn themselves into supermarkets for financial services, attempting to offer their customers "wall to wall" fulfillment of financial needs, others are narrowing their focus, shedding retail operations, or investment banking activities, concentrating on narrow market segments, such as active traders in the retail markets, for example. It appears that firms in the financial services industry choose their strategy in a trial and error mode, taking into account their existing portfolio of resources, lined up against evolving market opportunities and regulatory constraints.

Interestingly, as regulatory constraints in many countries have been considerably loosened and firms are free to extend their reach both geographically and in terms of product scope, managerial constraints are becoming more important. The difficulties in managing complex organizations with different corporate cultures across international borders increasingly become the crucial constraint on the expansion plans of financial services firms. In particular, specialized personnel and its care becomes a factor in determining the shape of institutions. Recent examples abound; they comprise the shifting of investment banking operations of banks in Continental Europe to London, where the environment seems to be more conducive for a business that requires the payment of "trucksize" bonuses in one year and brutal cuts in personnel the next. Alternatively, the painful abandonment of the merger of Dresdner Bank and Deutsche Bank due to the rebellion of mobile investment banking staffs in both institutions clearly highlights the constraints that impinge on remaking institutions, regardless of the apparent favorable economics of the total deal. Other examples include frequent failures of acquisitions of investment banking institutions by commercial banks, where value was destroyed by the inability of management to successfully integrate the different business cultures.

Another illustration of these crosscurrents is provided by the asset management industry.[5] On the one hand, one observes a distinct process of concentration resulting in the emergence of a large "fund factories," reaping economies of scale in processing, trading, and, most importantly, marketing and promotion. Only a large stable of funds provides the opportunity to wind up or merge poorly performing funds in order to enhance aggregate fund performance taking maximum advantage of the "survivor bias." At the same time, however, there will always be a significant group of investors looking for the fund manager with the magic hand who is able to outperform the market—for a while. This situation represents fertile ground for new investment management boutiques, hedgefunds, and other exotic specie.[6]

Such considerations point to a unique and perhaps defining characteristic of the market for financial services: it is incredibly segmented! To illustrate this point, while the sophisticated treasury operation of a multinational corporation,

---

5   For a comparative analysis of the drivers behind the changes of the mutual fund industry in Europe vs. the United States, see Walter (1999).

6   For a comprehensive survey of the asset management industry in the United States, see Wermers (2000).

staffed with academically trained analysts and savvy traders will buy the elements of financial contracts from a number of specialist suppliers and assemble them themselves into tailor-made packages, fitting the need of enterprise financial strategy, there exist many customers, including high networth professionals, who are not only willing to pay for advice but value the convenience of obtaining bundled products and one-stop shopping.

3.   Conclusions

a.   How Will Technology Affect the Banking Industry?

What does the presentation of dynamics described above, which focuses on the interaction between technology-driven changes in (a) the economics of producing and delivering financial services products and (b) the results from the regulatory arbitrage in globalized markets, tell us about the future of the financial services industry?

What will this mean for the "borders of banking," as reflected in the evolving structure of the banking industry?

I would venture first the conclusion that there will be little change in the core function: banks will continue to provide transactions and payment services due to their cost efficiencies and the fact that they have access to central bank clearing and overdraft facilities. In this respect alone, banks will continue to play a unique role at the core of any financial system. It is true they will have to fight for market share even in payment services with a number of institutions who get business through "preclearing" (i.e., credit cards and other payment providers). However, banks will be able to defend their core posititon, while markets "around the edges" of this core will be hotly contested.

The retailing industry is probably a good model to show the trends in the broader financial scene. Thus, it is probably safe to say that the trend toward the emergence of a limited number of financial supermarkets, selling their global reach, while being firmly entrenched in a "home" market will continue. Citibank Group, Deutsche Bank, HSBC, and Merrill Lynch represent a *type of institution* that will surely be around, although their names may change as mergers and acquisitions will be the major process in redefining the boundaries of such financial services firms.

Interspersed among these mammoth institutions will be a few quite specialized firms with global reach, primarily in the investment banking and asset management area, broadly defined. In today's environment, Goldman Sachs, Lehman, Fidelity, and Schwab would be the prominent examples. Further there will be room for institutions who build up strong positions in larger national markets, e.g., Japan and various economically integrated regions of the world. Among those, Development Bank of Singapore in South-Asia or Hypo-Vereinsbank in Continental Europe come to mind.[7] In between these more traditional structures of relatively large institutions, there will be lots of room for boutiques as well as chains of convenience stores. Their names will change rapidly in an M&A environment where "the fast eat the slow."

To sum up the industrial structure of the financial services firms, or "banks," if one prefers, will be characterized by dynamic change as the borders of individual firms are redefined rapidly by technologically driven changes in cost structures. The requirement for substantial reputational capital and remaining regulatory constraints will assure that this process will not be quite as rapid as technology permits.

b.   Challenges for the Regulatory Regime

All of this will force substantial change with regards to regulatory regimes. First of all, the forces pushing toward global coordination of regulatory regimes will increase in order to respond to the process of regulatory arbitrage described in the body of this paper. Second, there will be a new emphasis on regulatory technologies which are based on (a) increased disclosure, precariously balancing privacy concerns of financial firms and their clients with the public need for increased information, and (b) the continued and rapid substitution of direct supervision by a system that focuses on designing incentives for proper risk management in financial services institutions.

The direction of these trends is already clearly recognizable: they involve primarily appropriate-risk-based capital adequacy rules. The general introduction of layers of subordinated debt in banks which relies on market forces to obtain early warning signals represents another important step in this direction.

It would be naive to leave out of this catalogue some unresolved issues. To quote Hal Varian, "Technology changes—laws of economics do not." Applied to bank supervision, this means that the same old issues continue to be with us:

---

[7]   Note, I am not courageous or naive enough to make predictions about theses institutions per se. I simply use them as being representative of a type of institution, e.g., global players, regional universal banks.

- conflicts of interest
- moral hazard
- occasional irrationality of markets
- incomplete information

In the same category of unresolved issues are deposit insurance and public policy questions about "too big to fail—too small to be saved." It may well come to the point then that the rapidly changing shape of the financial services industry will compel fundamental reconsideration of the *scope* of regulation; possibly regulators will throw in the towel in frustration and move to the narrow bank concept where, in line with modern banking theory, they will segregate institutions that provide the crucial and unique function of providing liquidity to the rest of the participants in the economy. In turn, the remainder of the financial services industry will be left to market forces. At this point, economic theory may well clash irreconcilably with the real world of the political economy, with outcomes that are quite random.

Appendix

Financial Modernization: A Summary of the Gramm–Leach–Bliley Act of 1999[8]

On Friday, November 12, 1999, President Clinton signed into law the Gramm–Leach–Bliley Act of 1999 (henceforth GLBA), which repeals the Depression-era laws governing the U.S. financial system. Prior to the President's action, the bill cleared the Senate and House by substantial margins, 90–8 and 362–57, respectively, on November 4, 1999. In brief, GLBA is organized as follows: Tittle I, II, and III address organizational and regulatory structure issues, Title IV limits unitary thrift holding companies, Title V creates privacy protection, Title VI modernizes the Federal Home Loan Bank System, and Title VII addresses other issues. In this appendix, I discuss the bill, its components, and its implications on the financial industry landscape.

---

[8]   The Research Department, Federal Reserve Bank of Chicago has prepared this summary from legislation and press releases. It was published in *Capital Market News*, December 1999, which I have edited slightly.

Title I—Facilitating Affiliation between Banks, Securities Firms, and Insurance Companies

The main purpose of this initiative was to repeal the provisions of the 1933 Glass–Steagall Act and the 1956 Bank Holding Company Act, which both prohibit the affiliation of banking, securities, and insurance firms. The GLBA removes these barriers in various ways.

Bank holding companies will be allowed to enter the previously prohibited lines of business after qualifying as a financial holding company (FHC). FHCs will be allowed to engage in approved financial activities including insurance and securities underwriting and agency activities, merchant banking,[9] and insurance company portfolio investment activities. To qualify as a financial holding company all insured depository subsidiaries must have attained at least a "satisfactory" CRA rating at the time of application. The holding company will not be allowed to make new nonbanking acquisitions or engage in new financial activities if even one insured subsidiary falls below a CRA[10] rating of "satisfactory."

The GLBA also establishes guidelines, under which national banks may enter into new financial activities, including securities underwriting, through a financial subsidiary.[11] (National bank subsidiaries will not be permitted to engage in insurance underwriting, real estate investment and development, or merchant banking.[12]) The act also allows banks to directly deal in, underwrite, and purchase municipal bonds (including revenue bonds) for their own accounts. National banks must meet the following requirements in order to engage in these new activities:

- A bank and all of its insured depository institution affiliates must be well capitalized and well managed after the bank's investment in its financial subsidiaries is deducted from the bank's capital. A bank may not invest more than 45 percent of its assets, or $50 billion, whichever is less, in financial subsidiaries.

---

[9] Merchant banking is defined as the privately negotiated purchase of equity instruments by a financial institution, with the objective of selling the instruments at the end of an investment horizon, typically measured in years.

[10] CRA refers to the Community Reinvestment Act, a piece of legislation compelling U.S. banks to engage in social lending activities, especially the "inner cities."

[11] Financial subsidiaries are defined as any subsidiary other than those solely engaged in previously approved activities.

[12] In five years, national bank subsidiaries may be able to engage in merchant banking activities if both the Fed and the Treasury agree to allow it.

- Banks' loans to and investments in its subsidiaries would be limited to no more than 20 percent of the banks' capital. The bank and all depository affiliates must have at least a "satisfactory" CRA rating.
- The following ratings-based criteria must be met:
  - If a bank is among the 50 largest insured U.S. banks (in terms of assets), it must have at least one issue of long-term, unsecured debt outstanding that is rated within the top three rating categories of an independent rating agency.
  - If the bank is among the 100 largest insured U.S. banks (but not the 50 largest) the bank must meet either the rating requirement noted above or a comparable test jointly agreed upon by the Federal Reserve and the Treasury.
  - The above does not apply to national banks that are not among the 100 largest insured U.S. banks; that is, these institutions will not be able to engage in new financial activities.

The GLBA preempts state law, with certain exceptions, and puts national and state chartered banks on a more equal footing in exercising expanded powers. State banks will have to meet criteria similar to those a national bank must meet before they are able to establish new financial subsidiaries. The bill also mandates that all individuals engaged in insurance activities be appropriately licensed as required by state law.

The legislation provides for the streamlining of bank holding company supervision. The responsibility of the Federal Reserve (Board) is extended to the regulation of FHC organizations. In the Board's execution of its supervisory activities, it will accept reports from other regulatory agencies to the fullest extent possible. The Board may examine functionally regulated subsidiaries when they pose significant risk to affiliated banks and thrifts, when existing reports do not adequately depict risk monitoring systems, or where there is sufficient reason to believe that the subsidiary is not in compliance with federal law. The GLBA also contains provisions under which FHCs and banks may enter additional activities, and the Federal Reserve and Treasury must jointly approve these new activities.

Title II—Functional Regulation

The GLBA eliminated the broad broker-dealer exemption currently given to banks, with a few exemptions. This generally means that banks providing securities-related products would be subject to the same regulation as other providers. The Securities and Exchange Commission has primary regulatory authority over these activities; however, banks may continue to be participant in derivative activities involving credit and equity swaps. The SEC and Board share "rulemaking and resolution" powers regarding the treatment of new products that contain both banking and securities elements. The SEC is tasked with determining the treatment of any new hybrid product created by the industry. Prior to commencing the rulemaking process for any product, the SEC is required to seek agreement with the Federal Reserve regarding broker-dealer registration requirements.

In drafting the bill's language and Conference Report, lawmakers were careful to ensure that the SEC' oversight will not disturb traditional bank trust activities. To that end, the bill exempts banks that execute transactions in a trustee or fiduciary capacity from registration under federal security laws. There are two criteria that must be met to qualify for this exemption: The bank must be chiefly compensated for these services by means of administration of annual fees,[13] a percentage of assets under management per order processing fees,[14] or any combination thereof. Additionally, the institution may not publicly solicit brokerage business.

Title III—Insurance

The GLBA preserves and expands the primary jurisdiction of state insurance regulators over insurance activities, including serving as the functional regulator of insurance activities at national banks. Federal regulators are tasked with establishing consumer protection rules for the sale of insurance by national banks. They are to ensure, for instance, that customers are not mislead into believing such insurance products are federally insured, and are not coerced into buying other bank products.

Title IV—Unitary Savings and Loan Holding Companies

Lawmakers effectively closed the banking-commerce loophole created by the Unitary Thrift Charter. Existing unitary holding companies may only be sold to financial companies and de novo thrifts are prohibited from engaging in or affiliating with nonfinancial companies.

---

[13]   Administration or annual fees may be payable on a periodic basis (i.e., monthly or quarterly).

[14]   Processing fees may either be flat or capped and may not exceed the cost of executing transactions on behalf of customers.

Title V—Privacy (not related to issue at hand)

Title VI—Federal Home Loan Bank System Modernization (not related to issue at hand)

Title VII—Other Provisions (not related to issue at hand)

Bibliography

Berger, A.N., R. DeYoung, H. Genay, and G.F. Udell (1999). Globalization of Financial Institutions: Evidence from Cross-Border Banking Performance. Working paper WP-99-25, Federal Reserve Bank of Chicago.

Bhattacharya, S., and A. Thakor (1993). Contemporary Banking Theory. *Journal of Financial Intermediation* 3: 2–50.

Boot, A. (2000). Relationship Banking: What Do We Know? *Journal of Financial Intermediation* 9:7–25.

Boot, A., and A. Thakor (1996). Banking Structure and Financial Innovation. In A. Saunders and I. Walter (eds.), *Universal Banking: Financial System Design Reconsidered*. Homewood, Ill.: Irwin Professional Publishing.

Boot, A., and A. Thakor (1997). Financial System Architecture. *Review of Financial Studies* 10(3): 693–733.

Buch, C.M., and S.M. Golder (1999). *Foreign Competition and Disintermediation: No Threat to the German Banking System?* Kiel Working Paper 960, Institute of World Economics, Kiel.

Calem, P., and R. Rob (1999). The Impact of Capital-Based Regulation on Bank Risk-Taking. *Journal of Financial Intermediation* 8: 317–352.

Coleman, W.D. (1996). *Financial Services, Globalization and Domestic Policy Change*. New York: St. Martin's Press.

Corrigan, E.G. (1982). Are Banks Special?: A Revisitation. Federal Reserve Bank of Minneapolis, Annual Report 1982.

Corrigan, E.G. (2000). Are Banks Special?: A Revisitation. Federal Reserve Bank of Minneapolis, Special Issue 2000.

Degryse, H., and P. van Cayseele (2000). Relationship Lending Within a Bank-Based System: Evidence from European Small Business Data. *Journal of Financial Intermediation* 9: 90–109.

Deutsche Bundesbank (1997). *Monthly Report*, March. Frankfurt.

Deutsche Bundesbank (1999). *Monthly Report*, June. Frankfurt.

Deutsche Bundesbank (2000). *Banking Statistics*. Frankfurt.

Dufey, G. (1998a). Comments: Transformation of Banks and Bank Services. *Journal of Institutional and Theoretical Economics* 154(1): 137–143.

Dufey, G. (1998b). The Changing Role of Financial Intermediation in Europe. *International Journal of Business* (March): 49–67.

Dufey, G., and I.H. Giddy (1994). *The International Money Market.* Englewood Cliffs: Prentice-Hall Foundations of Finance.

Ferguson, R.W. (2000). Electronic Commerce, Banking, and Payments. Paper presented at the 36th Annual Conference on Bank Structure and Competition of the Federal Reserve Bank of Chicago, Chicago, May.

Financial Stability Forum Endorses Policy Actions Aimed at Reducing Global Financial Vulnerabilities (1999). Press Release and Exec. Summary from the Financial Stability Forum, Basle, Switzerland, March 26.

Franke, G. (1998). Transformation of Banks and Bank Services. *Journal of Institutional and Theoretical Economics* 154(1): 109–133.

Gensler, G. (2000). Technology Changing the World of Finance. Paper presented at the 36th Annual Conference on Bank Structure and Competition at the Federal Reserve Bank of Chicago.

Goodhart, C. (2000). Can Central Banking Survive the IT Revolution? *International Finance* 3(2):189–209.

Gopinath, D. (1999). Great Euro Expectations. *Institutional Investor* 33(4): 33.

Greenspan, A. (2000). Address given at the 36th Annual Conference on Bank Structure and Competition of the Federal Reserve Bank of Chicago.

Houpt, J.V. (1999). International Activities of U.S. Banks and in U.S. Banking Markets. *Federal Reserve Bulletin* (September): 600–615.

Kashyap, A.K., R. Rajan, and J.C. Stein (1999). Banks as Liquidity Providers: An Explanation for the Co-existence of Lending and Deposit-Taking. NBER Working Paper 6962, Cambridge, Mass.

Kroszner, R. (1998). On the Political Economy of Banking and Financial Regulatory Reform in Emerging Markets. *Research in Financial Services* 10: 33–51.

Kroszner, R., and P. Strahan (1999). What Drives Deregulation? Economics and Politics of the Relaxation of Bank Branching Restrictions. *Quarterly Journal of Economics* 114(4): 1437–1467.

Lehnert, A., and W. Passmore (1999). The Banking Industry and the Safety Net Subsidy. Working paper (rev. ed.), Federal Reserve Bank of Chicago.

Llewellyn, D.T. (2000). The New Economics of Banking. Paper presented at CIBI 2000 First International Conference on Innovation in the Banking Industry, March.

Meerschwam, D.M. (1991). *Breaking Financial Boundaries: Global Capital, National Deregulation, and Financial Services Firms.* Boston, Mass.: Harvard Business School Press.

Merrill Lynch and Co. (2000a). A Big Bang. April.

Merrill Lynch and Co. (2000b). Size and Structure of the World Bond Market: 2000. April.

Milbourn, T.T., W.A. Arnoud, and A.V. Thakor (1999). Megamergers and Expanded Scope: Theories of Bank Size and Activity Diversity. *Journal of Banking and Finance* 23: 195–214.

Morgan, D.P., and K.J. Stiroh (2000, May). Bond Market Discipline of Banks: Is the Market Tough Enough? Paper presented at the 36th Annual Conference on Bank Structure and Competition of the Federal Reserve Bank of Chicago, Chicago, May.

Padoa-Schioppa, T. (1999). Licensing Banks: Still Necessary? Unpublished paper, European Central Bank, Press Division, Frankfurt am Main.

Rajan, R.G. (1995). The Entry of Commercial Banks into the Securities Business: A Selective Survey of Theories and Evidence. Paper presented at the conference on Universal Banking, New York University.

Rajan, R.G. (1998). The Past and Future of Commercial Banking Viewed through an Incomplete Contract Lens. *Journal of Money, Credit, and Banking* 30(3): 524–550.

Schmidt, R.H., A. Hackethal, and M. Tyrell (1999). Disintermediation and the Role of Banks in Europe: An International Comparison. *Journal of Financial Intermediation* 9: 36–67.

Seth, R., D.E. Knoll, and S.K. Mohanty (1998). Do Banks Follow Their Customers Abroad? *Financial Markets, Institutions and Instruments* 7(4): 1–25.

Shull, B. (1999). The Separation of Banking and Commerce in the United States: An Examination of Principal Issues. *Journal of Financial Markets, Institutions and Instruments* 8(3): 1–55.

Spong, K., and R.J. Sullivan (1999). The Outlook for the U.S. Banking Industry: What Does the Experience of the 1980s and 1990s Tell Us? *Economic Review* (Federal Reserve Bank of Kansas City) 84(4): 65–83.

Thakor, A. (1996). The Design of Financial Systems: An Overview. *Journal of Banking and Finance* 20: 917–948.

Thakor, A., S. Bhattacharya, and A. Boot (1998). The Economics of Bank Regulation. *Journal of Money, Credit and Banking* 30(4): 745–770.

Wallison, P.J. (2000). The Gramm-Leach-Bliley Act Eliminated the Separation of Banking and Commerce: How This Will Affect the Future of the Safety Net. Paper presented at the 36th Annual Conference on Bank Structure and Competition of the Federal Reserve Bank of Chicago, Chicago, May.

Walter, I. (1997). Universal Banking: A Shareholder Value Perspective. *European Management Journal* 15(4): 344–360. Reprinted in *Finanzmarkt und Portfolio Management*, Switzerland, No. 1, 1997.

Walter, I. (1998). *Global Competition in Financial Services.* Cambridge, Mass.: Bollinger.

Walter, I. (1999). The Asset Management Industry in Europe: Competitive Structure and Performance Under EMU. In J. Dermine and P. Hillion (eds.), *European Capital Markets With a Single Currency.* Oxford: Oxford University Press.

Wermers, R. (2000). Mutual Fund Performance. *Journal of Finance* 35(4): 1655–1703.

Comment on Gunter Dufey

Ingo Walter

Gunter Dufey provides a very worthwhile discussion of the role of banks in the modern economy—focusing on issues of static and dynamic efficiency properties, risks associated with institutional and systemic failure, and regulatory requirements for optimizing the former within tolerable regulatory bounds to deal with the latter in what is by now a highly competitive global industry. He recognizes that many of the properties that made classic "banks" (institutions that accept deposits from the general public and make commercial loans) unique have fallen by the wayside, so that banking functionality in the narrow sense has increasingly been marginalized and banks as institutions have been forced into massive strategic adaptation, a process that is ongoing.

In this context, Gunter Dufey properly notes the role of technological change and innovation in financial products and processes, evolving public policy toward financial services firms, and the globalization of markets as well as firms. He implicitly recognizes the need to focus on financial intermediation *processes* as opposed to specific *institutional forms*, and hence the need for function-based, rather than institution-based, public policies and business strategies—a mismatch that has in the recent past led to periodic regulatory errors and strategic missteps.

Differences in financial intermediation processes and institutional design across time and space are striking. In the United States, "banks" using the above definition have seen their market share of domestic financial flows between end-users of the domestic financial system (i.e., between ultimate savers/investors and ultimate borrowers/issuers) decline from about 75 percent in the 1950s to under 25 percent today. In Europe (EU plus Switzerland), the change has been much less dramatic, and the share of financial flows running through the balance sheets of banks continues to be well over 60 percent—but declining nonetheless. And in Japan, banks continue to control in excess of 70 percent of financial intermediation flows. Most emerging market countries cluster at the highly intermediated end of the spectrum. Even here, there is plenty of evidence of declining market shares of traditional banking intermediaries. Classic banking functionality, in short, has been in long-term decline more or less worldwide.

Where has all the money gone? As Gunter Dufey notes, disintermediation and financial innovation has redirected financial flows through the financial markets. Ultimate savers increasingly use the fixed-income and equity markets directly or go through fiduciaries which, through vastly improved technology, are able to provide substantially the same functionality as classic bank accounts—immediate access to liquidity, transparency, safety, etc.—coupled to a higher rate of return. The one thing they cannot guarantee is settlement at par, which in the case of transactions balances (e.g., money market mutual funds) is mitigated by portfolio constraints mandating high-quality, short maturity financial instruments. Ultimate users of funds have benefitted from enhanced access to financial markets across a broad spectrum of maturity and credit quality using conventional and structured financial instruments. A broad spectrum of derivatives overlays the markets, making it possible to tailor financial products to the needs of end-users with increasing granularity, further expanding the availability and reducing the cost of financing, on the one hand, and promoting portfolio optimization, on the other. And as the end-users have themselves been forced to become more performance-oriented in the presence of much greater transparency and competitive pressures, it has become increasingly difficult to justify departures from highly disciplined financial behavior.

Two important and related differences are encountered in this generic financial-flow transformation: intermediation shifts (1) from book-value to market-value accounting and (2) from a more intensively regulated to a less intensively regulated channel, generally requiring less oversight and less capital. Both have clear implications for the efficiency properties of financial systems and for their transparency, safety, and soundness.

The end-result of this disintermediation, taking the United States as an example, is evident from the rough comparisons in Table 1. With about 28.9 percent of global GDP, U.S. banking assets and syndicated bank loans are well underweight (they are overweight in Europe and Japan), whereas both bond and stock market capitalizations, capital market new-issues, and fiduciary assets under manage-

Table 1: The U.S. Financial System in Perspective (percentage share of global)[a]

Population[b]	4.5	Completed M&A (by value)	52.8
GDP	28.9	Pension assets under mgt.	59.4
Banking assets	10.6	Mutual fund assets	53.0
Syndicated lending	13.5	Asset management (AUM)	51.1
Bond market capital	44.9	Loan lead-managers	77.2
Equity market capital	50.0	Debt & equity bookrunners	66.3
Non-govt. debt new issues	53.2	M&A advice (by transaction value)	78.6
Equity new issues	57.0		
[a]1999 data unless otherwise noted. — [b]Population data for 1995.			

ment are well overweight (they are underweight in Europe and Japan). One result is that U.S. financial intermediaries have come to dominate various intermediation roles—over half of global asset management mandates, over 77 percent of lead manager positions in wholesale lending, two-thirds of bookrunning mandates in global debt and equity originations, and almost 80 percent of advisory mandates (by value of deal) in completed global merger and acquisitions transactions. Indeed, it is estimated that, in 1999, U.S.-based investment banks captured about 70 percent of the fee income on European capital markets and corporate finance transactions (see Smith and Walter 2000).

Why? Table 2 suggests that the reasons include the size of the U.S. domestic market (roughly two-thirds of global capital-raising and M&A transactions in recent years), early deregulation of markets (but not of institutions) dating back to the mid-1970s, and performance pressure bearing on institutional investors, corporate and public-sector clients alike, leading to an undermining of client loyalty in favor of best price and best execution. Perhaps as an unintended consequence of separated banking from 1933 to 1999, institutions dominating disintermediated finance evolved from close-knit partnerships with unlimited liability to full-service investment banks under intense shareholder pressure to manage their risks and sweat their available capital in full knowledge that continued market discipline was likely to preclude regulatory bailouts.

Table 2: Factors behind U.S. Dominance in Global Financial Markets

• Size of home market.
• Early market deregulation & innovation-oriented market supervision.
• Rapid erosion of client loyalty.
• Strong performance orientation of independent broker-dealers, based on separated banking rules and partnership history.
• High capital requirements against securities positions.
• Strong risk management skills.
• Effective capital allocation by business-line.
• Absence of specific securities firm bailouts.

Given these ongoing developments, what about the implications for the industrial organization of the financial intermediation industry? Traditional banks have had to "go with the flow" and develop competitive fiduciary, underwriting, trading, and risk management capabilities under constant pressure from other banks and, most intensively, from other types of financial services firms. Gunter Dufey's paper contains a table (repeated in modified form in Table 3) which highlights one of the key industrial organization differences between Europe and the United States. Given the long history of universal banking, banks dominate most intermediation functions in Europe. Line-of-business limits in force since 1933 in the United States, on the other hand, have probably contributed to a much more heterogeneous structure in terms of financial intermediaries' institutional forms.[1]

Table 3: Structure of Financial Institutions[a]

United States	Europe[b]
Commercial banks	Banks
Savings institutions	Banks
Credit unions	Banks
Finance companies	Banks
Securities brokerage	Banks
On-line brokerage	Banks
Investment banks	Banks
Mutual fund companies	Mainly affiliates of banks
Mortgage companies	Affiliates of banks and mortgage banks
Insurance companies	Insurance companies & Allfinanz affiliates of banks

[a]Adapted from Gunter Dufey *The Blurring Borders of Banking*, this conference volume. — [b]Significant intercountry differences exist in European markets.

This is confirmed in Table 4. With financial intermediation distorted by regulation, a half-century ago, banks dominated classic banking functions, broker-dealers dominated capital market services, and insurance companies dominated most of the generic risk management functions. Mainly in savings products was there significant cross-penetration between types of financial intermediaries. Fifty years later, this functional segmentation has changed almost beyond recognition despite the fact that full *de jure* deregulation was not implemented until the end of the period—with the Gramm–Leach–Bliley Act of 1999 (GLBA). Not only has there been a virtual doubling of "strategic groups" competing for financial intermediation functions, but there is now massive cross-penetration among them. Most financial services can be obtained in one form or another from virtually every strategic group, each of which in turn is involved in a broad array of

---

[1]   For a discussion of the financial markets aspects, see Dermine and Hillion (2000) and Walter and Smith (2000).

Table 4: The U.S. Financial Services Sector in 1950 and 2000

Function \ Institution	Payment services	Savings products	Fiduciary services	Lending to		Underwriting issuance of		Insurance and risk management products
				business	consumer	equity	debt	
Banks	● ▲	● ▲	● ▲	● ▲	● ▲	●	●	●
Insurance companies	●	● ▲	●	● △	●	●	●	● ▲
Finance companies	●	●	●	● △	● ▲	○	○	●
Securities firms	●	● ▲	● ▲	●	●	● ▲	● ▲	●
Pension funds		● ▲	●	●	●			
Mutual funds	●	● ▲	●		●			
Diversified financial firms	●	●	●	●	●	○		●
Specialist firms	●	●	●	●	●	●	●	●

Note: ▲ denotes a major involvement, △ a minor one in 1950, whereas ● denotes a major involvement, ○ a minor one in 2000.

financial services. If cross-competition among strategic groups promotes both static and dynamic efficiencies, then the U.S. structure as it has evolved (partly as an unintended consequence of line-of-business regulation) probably serves macroeconomic objectives—particularly growth and economic restructuring—very well indeed.

Table 5 suggests that technological change, particularly leveraging the properties of the Internet, will further enhance financial intermediation efficiencies. It has already dramatically cut information and transaction costs for both retail and wholesale end-users as well as for the intermediaries themselves. The examples of on-line banking, insurance, and retail brokerage given in Table 5 are well known and continue to evolve and change the nature of the process, sometimes turning the prevailing business model on its head—e.g., financial intermediaries have traditionally charged for transactions and provided advice almost for free, but increasingly are forced to provide transactions services almost for free and to charge for advice. At the same time, on-line distribution of financial instruments such as commercial paper, equities, and bonds not only cuts the cost of market access but also improves and deepens the distribution and bookbuilding process—including providing issuers with information on the investor base. And it is only one further step to cutting out the intermediary altogether by putting the issuer and the investor or fiduciary into direct electronic contact. The same is true in secondary markets, with an increasing array of alliance-based competitive bidding utilities (FXall) and reverse auctions (Currenex.com) in foreign exchange and other financial instruments as well as interdealer brokerage, cross-matching and electronic communications networks (ECNs). This Internet-based technology overlay is likely to turbocharge the cross-penetration story depicted in Table 4.

A further development consists of automated end-user platforms such as CFOWeb.com for corporate treasury operations and Quicken 2000 for households, with real-time downloading of financial positions, risk profiles, market information, research, etc. By allowing end-users to "cross-buy" financial services from best-in-class vendors, such utilities may well up-end conventional thinking, which focuses on cross-selling, notably at the retail end of the end-user spectrum. If this is correct, financial firms that are following Allfinanz or bancassurance strategies may end up trapped in the wrong business model.

Table 6 is a rough sketch of financial intermediation in Europe, comparable to Table 4 for the United States, involving a dominant position across all functional areas by universal banking organizations and substantially less cross-penetration by more narrowly focused and specialist firms—and those that do exist in many cases are U.S.-based. It is interesting to speculate what the European matrix (especially within the euro-zone) will look like in ten years' time. Many observers would argue that a rich array of players stretching across a broad spectrum of strategic groups will serve Europe better than a financial monoculture based

Table 5: E-Applications in Financial Services (June 2000)[a]

Banking:
• On-line banking (CS Group, Bank-24, E*Trade Bank, Amex Membership B@nking, Egg)
Insurance:
• eCoverage (P&C)
Brokerage:
• E-brokerage (Merrill Lynch, MSDW, Fidelity, Schwab, E*trade, DLJ Direct, Consors)
Primary markets:
• E-based CP & bond distribution (UBS Warburg, Goldman Sachs)
• E-based direct issuance
• Governments (TreasuryDirect, World Bank)[b]
• Municipals (Bloomberg Municipal, MuniAuction, Parity)
• Corporates (CapitaLink[c], Intervest)
• IPOs (Hambrecht, Wit Capital, Schwab, E*Trade)
Secondary markets:
• Forex (Currenex.com, FXall.com)[d]
• Governments (Bloomberg Bond Trader, QV Trading Systems, TradeWeb EuroMTS)
• Municipals (QV Trading Systems, Variable Rate Trading System)
• Corporates (QV Trading Systems)
• Government debt cross-matching (Automated Bond System, Bond Connect, BondNet)
• Municipal debt cross-matching (Automated Bond System)
• Corporate debt cross-matching (Automated Bond System, Bond Connect, Bondlink, BondNet Limitrader)
• Debt interdealer brokerage (Brokertec, Primex, Equities – ECNs (Instinet, Island, Redi-Book, B-Trade, Brut, Archipelago, Strike, Eclipse)
• Equities cross-matching (Barclays Global Investors, Optimark)
Corporate finance end-user platforms (CFOWeb.com)[e]
Institutional investor utilities
Household finance utilities (Quicken 2000)

[a]Examples of leading players in each category, some not live as of 15 June 2000. — [b]Web-based bookbuilding u/w by 8 banks, Goldman Sachs bookrunner. — [c]Defunct. — [d]FXall = multibank site. Currenex = reverse currency auction. — [e]Includes OTC securities and derivatives through participating firms, information (MoneyLine Network), risk mgt. (Algorithmics), financial advice (Deloitte & Touche).

on massive universal banking organizations. At the same time, there is the question of what the U.S. system will look like in a decade, now that the artificial barriers between banking, insurance, and securities business carried out in the same organization have been swept away. So far, the most valuable financial services franchises in the United States and Europe in terms of market capitaliza-

Table 6: The European Financial Services Sector in 2000[a]

Function / Institution	Payment services	Savings products	Fiduciary services	Lending to business	Lending to consumer	Underwriting issuance of equity	Underwriting issuance of debt	Insurance and risk management products
Banks	●	●	●	●	●	●	●	●
Insurance companies		●	●					●
Finance companies				●	●			●
Securities firms[b]			●			●	●	●
Pension funds			●					
Mutual funds[c]		●	●					
Diversified financial firms		○	○	○	○			○
Specialist firms		●	●	○	●			●

[a]Major intercountry differences exist. — [b]Virtually all US-based. — [c]Most bank-affiliated or US-based. — *Note:* ● denotes a major involvement, ○ a minor one.

tion seem far removed from a financial intermediation monoculture—see Table 7. In fact, each presents a rich mixture of banks, asset managers, insurance companies, and specialized players.

As Gunter Dufey's paper suggests, the institutional microstructure of the financial intermediation process is a critical determinant of economic develop-

Table 7: The World's Most Valuable Financial Services Firms (regional top 20, July 2000, billions of dollars)[a]

North America		Europe	
Citigroup	150.9	HSBC Holding	98.7
AIG	141.6	Lloyds TSB Group	72.0
GECS[b]	139.9	Allianz	66.1
Bank of America	112.9	UBS AG	63.4
Berkshire Hathaway	109.4	ING Groep	50.8
Banc One	66.8	Zurich Fin. Serv.	49.1
Wells Fargo	66.4	Crédit Suisse Grp.	48.1
Chase Manhattan	61.1	Aegon	47.9
Morgan Stanley DW	55.1	Barclays Bank	45.8
American Express	54.5	Groupe AXA	41.4
First Union	44.7	NatWest	38.8
Charles Schwab	42.9	Generali	36.6
Goldman Sachs Group	32.3	Fortis	35.9
Merrill Lynch	30.2	Deutsche Bank AG	32.2
Allstate	29.5	Halifax	31.4
Associates First Capital	28.5	BBV	30.1
Bank of New York	27.2	Münchener Rück	29.8
J.P. Morgan	24.5	Abbey National	29.3
U.S. Bancorp	23.6	Swiss Re	27.6
Washington Mutual	22.7	Prudential Corp.	25.7
Total top 20	1,189.8	Total top 20	877.0

[a]Top Asia market capitalization rankings are DKB-Fuji-IBJ 86.1, Bank of Tokyo-Mitsubishi 61.8, Sakura-Sumitomo Bank 77.8, Sanwa-Tokai-Asahi 59.5. — [b]Estimated. GECS net 1998 after-tax earnings contribution as a percent of General Electric Company consolidated after-tax earnings times GE's July 1999 market capitalization. Data: *Business Week*, 12 July 2000; Bloomberg Financial Markets (Website); *American Banker*, 16 August 2000.

ment, and, despite financial globalization and the gradual suppression of geo-graphic proximity as a constraint, financial intermediation microstructure at the local and regional level continues to be of vital importance. So far, the evolution of the financial systems in a competitive environment provides plenty of reasons for optimism.

Bibliography

Dermine, J., and P. Hillion (eds.) (1999). *European Capital Markets With a Single Currency*. Oxford: Oxford University Press.

Smith, R.C., and I. Walter (2000). Global Wholesale Finance: Structure, Conduct, Performance. Paper presented at the 22nd Annual Colloquium of the Société Universitaire Européenne de Recherches Financières (SUERF), Vienna, 27–29 April.

Walter, I., and R.C. Smith (2000). *High Finance in the Eurozone*. London: Financial Times – Prentice Hall.

Marco Becht and Colin Mayer

Corporate Control in Europe

632

634

1.   Existing Studies of Ownership and Control

The classic study of the ownership and control of corporations is Berle and Means (1932). On the basis of an analysis of U.S. corporations, they concluded that

> the separation of ownership from control has become effective—a large body of security holders has been created who exercise virtually no control over the wealth which they or their predecessors in interest have contributed to the enterprise.... The separation of ownership from control produces a condition where the interests of owner and of ultimate manager may, and often do, diverge, and where many of the checks which formerly operated to limit the use of power disappear. (Pages 6 and 7)

Berle and Means's analysis stimulated a huge literature on the control problems created by the separation of ownership and control. Their arguments were amplified in "the corporation problem" literature of the 1960s. Manne (1965) argued that the market, not management, controls the "modern corporation" but Grossman and Hart (1980) cast doubt on the effectiveness of the takeover mechanism. Pessimism culminated in Jensen's (1989) prediction of "The Eclipse of the Public Corporation" and a move to private corporations with high levels of debt. Re-

Remark: This paper is an abridged version of the introduction to a forthcoming book entitled *The Control of Corporate Europe*, which reports the results of an international study of corporate control in Europe undertaken by research members of the European Corporate Governance Network: Jonas Agnblad, Erik Berglöf, Marcello Bianchi, Magda Bianco, Laurence Bloch, Ekkehart Böhmer, Ariane Chapelle, Rafel Crespí-Cladera, Abe de Jong, Luca Enriques, Miguel A. García-Cestona, Marc Goergen, Klaus Gugler, Peter Högfeldt, Rezaul Kabir, Susanne Kalss, Elizabeth Kremp, Teye Marra, Luc Renneboog, Ailsa Röell, Alex Stomper, Helena Svancar, and Josef Zechner. This paper draws on the results of the collective research of the Network. We are grateful to participants at the conference for very helpful comments.

cently, the U.S. public corporation, with apparently strong outside directors, shareholder activists, and strong legal protection of shareholders, is again the global favorite. But the latest evidence suggests that managers are as powerful as ever. Weak boards with ample formal power, insurmountable takeover barriers, and anti-blockholder regulations continue to prevent owners from exerting control.

To date, very little is known about the control of corporations outside the United States. In fact, such has been the influence of Berle and Means that the textbook description of dispersed ownership and separation of ownership and control has been presumed to be universally applicable. But over the last few years, evidence has emerged that has questioned this view.

Franks and Mayer (1995) described two types of ownership and control structures—what they termed the "insider and outsider" systems. The outsider system corresponds to the Berle and Means description of the United States—ownership is dispersed amongst a large number of outside investors. Both the United Kingdom and United States have outsider systems. In the United Kingdom, a majority of equity is held by financial institutions, predominantly pension funds and life assurance companies. In the United States, individual shareholders are more widespread. But in neither country do institutions or individuals hold a large fraction of shares in a company. As a consequence, they exert little direct control over corporations, and the separation of ownership and control described by Berle and Means is observed.

However, Franks and Mayer also noted a quite different system that existed in Continental Europe. There, few companies are listed on stock markets and those companies that are listed have a remarkably high level of concentration of ownership. Franks and Mayer observed that in more than 80 percent of the largest 170 companies listed on stock markets in France and Germany, there is a single shareholder owning more than 25 percent of shares. In more than 50 percent of companies, there is a single majority shareholder. The corresponding figures for the United Kingdom were 16 percent of largest 170 listed companies had single shareholders owning more than 25 percent of shares and 6 percent had single majority shareholders. Concentration of ownership is staggeringly high on the Continent in comparison with either the United Kingdom or United States.

Franks and Mayer noted that the ownership of Continental European companies is primarily concentrated in the hands of two groups: families and other companies. Cross-shareholdings and complex webs of intercorporate shareholdings are commonplace in some countries. Companies frequently hold shares in each other in the form of pyramids by which company A holds shares of company B, which holds shares of company C, etc. They also observed that bank ownership of corporate equity was generally quite modest, despite the attention that has been devoted to the role of bank shareholdings in cementing bank-firm

relations. In some, but by no means all, Continental European countries, ownership by the state is appreciable. Barca et al. (1994) report similar results for Italy.

La Porta et al. (1997, 1999) have recently extended Franks and Mayer's study
to many more countries. They have found that the observation on insider systems
which Franks and Mayer made about Continental European countries applies
widely around the world. They conclude that the Berle and Means corporation is
much less applicable than previously thought. Instead, the insider system appears
to dominate.

Interesting though these studies are, they both have serious methodological
problems. Firstly, the coverage of the studies is very limited. The Franks and
Mayer analysis refers to the largest 170 firms in France, Germany, and the United
Kingdom. The La Porta et al. study (1999) is restricted to the largest 20 firms in
each of their 27 countries.[1] Franks and Mayer have a reasonably large number of
large corporations, but in a very small number of countries; La Porta et al. have a
very small number of companies in a large number of countries.

Secondly, the analysis of control in both papers is rudimentary. In fact, though
both papers refer to ownership, what they actually measure is voting control.
Data on ownership are in general simply not available and all that can be measured is voting rights. More significantly, an analysis of control of corporations is
complex. In a subsequent paper on Germany, Franks and Mayer (1995) noted
that shareholdings are frequently in the form of pyramids in which the control
that the owner at the top of the pyramid exerts can be disproportionate to the cost
of their investments as measured by their cash flow rights. In order to identify
where control resides, it is necessary to trace it back up through the pyramid to
the ultimate shareholders at the top of the pyramid. La Porta et al. (1999) attempt
to do this but their data cannot capture the full complexity of control arrangements that exist in corporate Europe, as we describe below.

Over the last few years, the possibility of undertaking a much more precise
analysis of the control of European corporations has emerged. This has arisen as
a consequence of vastly improved disclosure standards in Europe. To facilitate
the creation of a truly European equity market, the European Union has adopted
the Large Holdings Directive (88/627/EEC). This Directive has created the
unique opportunity to study corporate control. Access to these data has been improved by cross-listings and integration of capital markets. Companies with a
foreign listing in the United States have to file a Form 20-F that includes a spe-

---

[1]   It is unclear whether the La Porta et al. study refers to the largest 20 companies in
each of their countries or the largest 20 companies in Worldscope, which is their
main data source. Since Worldscope coverage is far from comprehensive, the 20
companies do not necessarily correspond with the largest 20 in any one country.

cial section on the "Control of the Registrant." Voting blocks of 10 percent or larger that have been reported to the companies as a result of the European directive must be disclosed. In future, it will also be possible to perform this type of analysis in Eastern Europe, since before they can join, the accession countries of the European Union will have to transpose the European Union's directives. As a result, better information on voting power concentration will become available for these countries as well. Internationally, the OECD Principles of Corporate Governance and related efforts by IOSCO are likely to make disclosure more effective and to provide further research opportunities.

Until a few years ago, international comparisons of financial systems focused on the financing of firms and, in particular, the role of banks in funding companies. Distinctions were drawn between supposedly bank-oriented financial systems, such as Germany and Japan, and the market-oriented systems of the United Kingdom and United States. But closer analysis revealed the fragility of these distinctions. Mayer (1988), Edwards and Fischer (1994), and Corbett and Jenkinson (1996) noted that the amount of lending coming from German banks has been modest over a long period of time. Edwards and Fischer (1994) argued that there is little support for the conventional wisdom that German banks are actively involved in monitoring and controlling corporations.

In the case of East Asia, in particular Japan, research focused on the relative performance of *keiretsu* and non-*keiretsu* companies (Aoki 1990; Prowse 1992; Kaplan and Minton 1994). Initially, it appeared that the closer bank-firm relations in *keiretsu* groups are reflected in fewer credit constraints and the provision of more financing during periods when companies are in financial difficulty (Hoshi et al. 1991). However, even this view has been questioned over the last few years. Kang and Stulz (2000) reported that bank-dependent firms suffered significantly larger wealth losses and invested less than other firms during 1990 to 1993 when the Japanese stock market dropped appreciably. Weinstein and Yafeh (1998) recorded that close bank-firm ties increased the availability of capital to Japanese firms but did not lead to higher profitability or growth because of banks' market power.

The distinction between bank- and market-oriented financial systems is therefore fragile. In contrast, the differences in ownership and control of corporations noted above are pronounced. This raises two questions. Firstly, what are their causes and, secondly, what are their consequences. We examine existing debates on these two issues in Section 2. In Section 3 we summarize the results of the European Corporate Governance Network study and in Section 4 we conclude the paper.

2.   The Causes and Consequences of Ownership Concentrations

Over the last few years, much attention has been given to the influence of regulation and legal form on corporate ownership and control. Black (1990) and Roe (1994) pioneered the "overregulation" thesis that has been a dominant force in the debate. Roe (1994) documents that regulation prevents potentially important investors from holding blocks. Black (1990) argues that regulation makes it costly to hold blocks and causes shareholder passivity in the United States.

If the separation of ownership and control is induced by regulation and is not efficient, why was the regulation introduced and why is it not repealed? Roe argues that the rise of the Berle and Means corporation in the United States was not simply a response to the forces of economic efficiency, but a reflection of populist politics. Concerns about concentration of control in particular in the hands of such banks as J.P. Morgan led to a backlash in the imposition of regulation restricting the involvement of banks in corporate activities. Dispersed ownership therefore resulted from the introduction of regulatory impediments to concentrations in ownership prompted by a populist political agenda. Bebchuk and Roe (1999) argue that inefficient regulation can persist as a result of "path-dependence."

The overregulation argument views the difference between the United States and other countries as reflecting impediments to the free choice of corporate structure in the United States. However, over the last few years an exactly contrary view has been presented. Far from U.S. corporations being impeded by regulation from choosing appropriate structures, "underregulation" and, in particular, weak investor protection undermine the financing of firms in most countries of the world. According to La Porta et al. (1997), concentrations of ownership and complex control vehicles are a response to inadequate protection of investors. Faced with a risk of exploitation by self-interested managers, investors require powerful mechanisms for exercising control and they do so through holding large ownership stakes in companies and exerting voting power that is disproportionate to the amount that they invest in firms. This argument has been formalized by Bebchuk (1999), who presents a rent-seeking theory of the evolutions of ownership and control and voting power leverage. If blockholder control is induced by underregulation and is not efficient, why has no regulation been introduced? La Porta et al. (2000) invoke similar arguments to those of Roe (1994) and Bebchuk and Roe (1999): rent-seeking blockholders are powerful and have prevented stock exchanges and regulators from curbing their power.[2]

---

[2]   In the political theory of Roe (1991), U.S. managers influence regulation to obtain or protect rents.

Easterbrook and Fischel (1991) adopt a third position. They argue that the corporate structures we observe have an economic purpose. The Berle–Means corporation and the closely held corporation are efficient, but in different contexts. If they were not, they would not have grown and survived. Easterbrook (1997) argues that the structure and needs of the financial system and the forces of the market create the necessary regulation, not the other way round.

Roe views the U.S. corporation as being weighed down by regulatory restrictions imposed by an earlier political agenda; La Porta et al. (1997) argue that strong investor protection in common law countries and underregulation elsewhere has allowed external financing to occur on a larger scale under common law. Easterbrook (1997) argues that the size of a country's stockmarket, as determined, for example, by its pension system, determines regulation. For Roe, the regulatory barriers to, for example, bank participation in corporate ownership should be broken down. For La Porta et al., stronger investor protection is required in most countries of the world. For Easterbrook and Fischel, regulation adapts to the needs of the market.

In this paper, we show that regulation visibly affects the relationship between ownership and control. We consider how the control of dominant investors changes as outside ownership is brought in. Dual class shares, nonvoting shares and pyramids allow dominant investors to exert disproportionate degrees of voting power as outside ownership comes in. This is termed "a private control bias." Rules preventing or discouraging shareholders from exercising voting rights, in particular in takeovers or board elections, lead to a dispersion of voting power. This can either create a market or a management control bias. In the absence of anti-takeover devices, dispersed voting power can be exercised through a hostile bid; Manne's (1965) market for corporate control. If the bidder holds a minority stake of sufficient size or has other incentives to launch a bid, minority investors can free ride on the control exerted by others and thereby derive control disproportionate to their investment. However, rules that discourage voting blocks can also lead to a "management bias." In most countries a variety of formal and informal devices shielding management from external control exist (poison pills, voting caps, weak and captured boards). Protecting minorities by curbing the voting power of the blockholder can severely tilt the balance of cash-flow rights and power in favor of management.

Figure 1 illustrates. It shows that rules protecting dominant investors push the line linking ownership to control upwards to the left. Control therefore declines less than proportionately with reduction in ownership. On the other hand, rules preventing or discouraging shareholders from exercising voting rights push the ownership control line down to the right. Control then declines more than proportionately with ownership.

Figure 1: Private Control Bias and Management or Market Control Bias

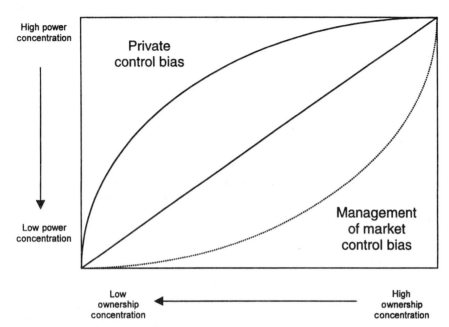

Note: Power increases proportionally with ownership under one-share-one-vote, in the absence of pre- and post-bid anti-takeover devices and under "neutral" regulation. Rules and devices protecting dominant investors push the line linking ownership to control upwards to the left. Control therefore declines less than proportionately initially with reduction in ownership. On the other hand, strong minority protection pushes the ownership-control line down to the right. Control then declines more than proportionately with ownership. In the absence of anti-takeover devices such as poison pills or voting caps, this will result in a market control bias. More likely, it results in a management control bias.

There is a similar divergence of view about the consequences of ownership concentration. The most popular view of corporate governance is the agency one. Managers run firms according to their own interests and agenda when the company is widely held. As far as possible, incentives are used to align the interests of managers and shareholders, but active monitoring and control of companies is also required. But, as noted by Berle and Means, in the presence of dispersed ownership, investors have little incentive to engage actively in this and instead prefer to "free ride" on the monitoring performed by others. Some ownership concentration is therefore required to ameliorate managerial agency problems (Berle 1958).

But while concentrations of ownership may reduce agency problems, they create a second conflict. The interests of holders of large blocks of shares may diverge from those of minority investors. Large shareholders are in a position to engage in activities that benefit them at the expense of minority investors. In Jensen and Meckling (1976) the owner-manager can freely choose the level of private benefit consumption because he/she retains all control rights while selling only cash-flow rights. The most obvious example of such "private benefits" is shifting assets and corporate value through transfer pricing between companies in which large shareholders have an interest. In some countries there are explicit restrictions on these activities, requiring for example, that parent and subsidiary act on an arms-length basis and use market testing to price transactions. In other countries, there are few restrictions.

While the "overregulation" literature sees weak owners confronted with strong managers in countries with dispersed ownership, the rent-seeking literature argues that strong owners in countries with concentrated ownership and/or control exploit weak minorities. In particular, La Porta et al. (1999) have argued that the latter is the more serious problem confronting most countries and the emergence of ownership concentrations in the face of weak investor protection is a deficiency, not an advantage, of these systems.

Carlin and Mayer (2000a, 2000b) and Mayer (2000) argue that there is no one dominant system. Different types of ownership and control are suited to different types of activities. They argue that concentrated ownership benefits activities that require long-term, committed investors. Dispersed ownership benefits short-term investments that require greater flexibility and less commitment. In particular, Mayer (2000) argues that there is a need to match periods for which control can be expected to be retained (what is termed the "influence period") with the "realization period" of projects (the time taken for projects to come to fruition). Too short an influence relative to a realization period leads to rejection of long-term investments. Too long an influence relative to realization period leads to inefficient retention of control. Dominant owners are able to retain control over long periods, whereas managers facing markets in corporate control with dispersed ownership are subject to short influence periods. The latter therefore promote efficient transfers in control for investments with short realization periods, whereas the former encourage investments in activities with long realization periods.

3.   Results

Traditionally, European corporations and their shareholders value privacy. In Austria, Belgium, Denmark, France, Germany, Luxembourg, the Netherlands, Portugal, and Spain most listed companies issue bearer shares. By definition, ownership and voting power are hard to trace. However, this can also be difficult in countries with registered shares (Finland, Greece, Ireland, Italy, Sweden, and the United Kingdom). Company law and securities regulations do not always force companies to provide access to share registers; alternatively, shares might be held in nominee accounts ("street names"). Even when the names of the beneficial owners are known, the share register provides no information on voting pacts and other arrangements that tie individual stakes into blocks of votes cast by the same person or entity.

The comparative analysis in this paper relies on a new disclosure standard that partially overcomes these problems. It applies throughout the European Union and provides for the disclosure of 10%+ (often 5%+) *voting blocks.*

a.   Average Size of Voting Blocks

Table 1 reports the proportion of votes controlled by the largest voting block in eight countries. In 50 percent of nonfinancial listed companies in Austria, Belgium, Germany, and Italy, a single blockholder (an individual investor or group of investors) controls more than 50 percent of voting rights. In 50 percent of Dutch, Spanish, and Swedish companies, more than 43.5 percent, 34.5 percent, and 34.9 percent, respectively, of votes are controlled by a single blockholder. In contrast, the median blockholder in the United Kingdom controls only 9.9 percent of votes and in the United States the median size of blockholding of companies quoted on both Nasdaq and NYSE is just above the disclosure level of 5 percent (8.6 percent and 5.4 percent).

The picture looks very different when one goes down to the second and third largest blocks (Table 1). The median size of the second largest voting block is 2.5 percent in Austria, 10.2 percent in Belgium, 5.9 percent in France, 7.6 percent in Italy, and 8.7 percent in Sweden. In Germany it is below the disclosure level. The median size of the third largest voting block is 4.7 percent in Belgium, 3.4 percent in France, 3.0 percent in Italy, and 4.8 percent in Sweden. In both Austria and Germany it is below the disclosure level. The size of voting blocks therefore decreases rapidly beyond the largest shareholder. Voting power is concentrated on the Continent not only because of the existence of large blockholders but also because of the absence of other voting blocks.

Table 1: Size of Ultimate Voting Blocks by Rank[a]

Country	No. of comp.	Largest voting block Median	2nd Largest voting block Median	3rd Largest voting block Median
Austria [1]	50	52.0	2.5	0.0[b]
Belgium [2]	140	56.0	6.3	4.7
Germany [3]	372	57.0	0[b]	0[b]
Spain [4]	193	34.5	8.9	1.8
France [5]	CAC40	20.0	5.9	3.4
Italy [6]	214	54.5	5.0	2.7
Netherlands [7]	137	43.5	7.7	0[b]
Sweden [8]	304	34.9	8.7	4.8
United Kingdom [9]	207	9.9	6.6	5.2
United States [10]				
NYSE	1309	5.4	0[b]	0[b]
Nasdaq	2831	8.6	0[b]	0[b]

[a]The table reports the size of the largest, 2nd largest, and 3rd largest median voting block for nonfinancial companies listed on an official market. For France only the main stock price index (CAC40) is covered. — [b]There is no 5 percent + voting block.

Source: [1] Gugler et al. (2000: Figure 1, data for 1996); [2] computed by Becht, data for 1995; [3] Becht and Böhmer (2000: Figure 2, data for 1996); [4] Crespí-Cladera and García-Cestona (2001, data for 1995); [5] Bloch and Kremp (2001, data for 1996); [6] Bianchi et al. (2000: Table 6, data for 1996); [7] De Jong et al. (2001: Table 4, data for 1996); [8] Agnblad et al. (2001: Table III A, data for 1998); [9] Goergen and Renneboog (2001, data for 1992); excludes blocks held by directors; [10] Becht (2001), includes blocks held by directors and officers; data for 1996.

In the United Kingdom, the median size of the second largest block is 7.3 percent, and of the third largest block is 5.2 percent. The size of the largest block is therefore appreciably smaller in the United Kingdom than on the Continent but the size of blocks does not decline very rapidly thereafter. Indeed, the third largest block and beyond is larger in the United Kingdom than any other country in this study. Even beyond the tenth largest block holding, the mean voting block in the United Kingdom is greater than 3 percent, whereas it is below disclosure levels in virtually all the Continental European countries in this study. On the Continent, the largest blockholder exerts dominant voting control in relation to other blockholders. In the United Kingdom, no individual blockholder in general exerts dominant control; instead it can only come from coalitions of investors. The situation is the same in the United States but the potential coalition to exert effective control must be even larger than in the United Kingdom.

Voting power is much more concentrated on the Continent than in the United Kingdom or the United States. Coalitions between shareholders are required in the United Kingdom and United States to exercise control. This points to a private control bias on the Continent and a management or market control bias in the United Kingdom and United States. The next section will provide support for this from an analysis of distributions of voting control in different countries.

b.   Distributions of Voting Control

Figures 2 to 4 show the cumulative distributions of largest voting blocks (from smallest to largest) in listed companies in seven European countries and for NYSE and Nasdaq companies in the United States. The figures report the fraction of companies in a country with largest blocks less than the values reported on the vertical axis. A cumulative distribution close to the 45° line reflects a uni-

Figure 2: Percentile Plot of Largest Voting Blocks in Listed Firms of Various Countries

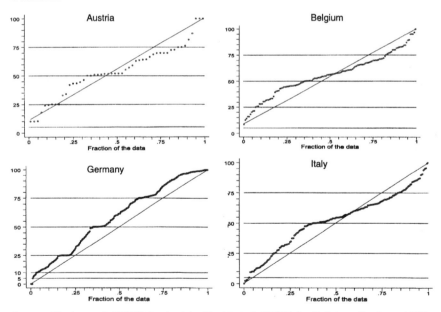

Source: Gugler et al. (2001) for Austria, Becht et al. (2001) for Belgium, Becht and Böhmer (2001) for Germany, Bianchi et al. (2001) for Italy.

Figure 3: Percentile Plot of Largest Voting Blocks in Listed Firms of Various Countries

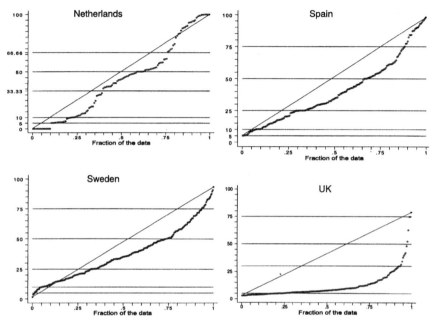

Source: De Jong et al. (2001) for the Netherlands, Crespi-Cladera and García-Cestona (2001) for Spain, Agnblad et al. (2001) for Sweden, Goergen and Renneboog (2001) for the United Kingdom.

Figure 4: Percentile Plot of Largest Voting Blocks in U.S. Companies Listed on the NYSE (left panel) and Nasdaq (right panel)

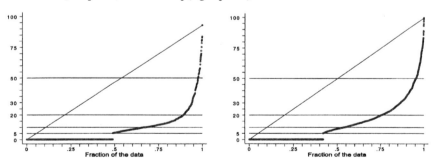

Source: Becht (2001).

form density of firms by voting blocks. A distribution above the 45° line reflects a preponderance of large voting block companies and a distribution below the 45° line indicates a large amount of dispersed voting control.

In Germany, the cumulative distribution is above the 45° line; in Austria, Belgium, and Italy, the distributions are close to the 45° line. In the Netherlands and Spain, they are below the 45° line. But the real contrast is between the United Kingdom and United States on the one hand and Continental Europe on the other. The distributions for firms in the United Kingdom and United States (both NYSE and Nasdaq firms) are very similar. In the United States, a majority of shareholdings are below the disclosure level of 5 percent and there are very few majority voting block companies in either country. The United Kingdom and United States therefore display pronounced market/management control bias; the Netherlands and Spain weak management control bias; the Austrian, Belgian, and Italian distributions are quite neutral and Germany displays a private control bias.

Still more interesting are the concentrations of firms on the distributions. In Austria and Germany, there is clear bunching of firms around 25 percent, 50 percent, and 75 percent voting blocks. These correspond to significant voting levels in both countries (blocking minority, majority and super-majority voting). In Belgium and Italy, there are concentrations just in excess of 50 percent voting blocks. In the United Kingdom, there are few blocks in excess of 30 percent. This corresponds to the level at which mandatory bids have to be made for all the shares of a target company. Takeover rules have therefore discouraged the accumulation of share blocks in excess of 30 percent in the United Kingdom. In the United States, shareholdings in excess of 10 percent and 20 percent may have undesirable regulatory control implications (in terms of disposal of shares and liabilities for federal law violations; see Black 1990). This is consistent with the concentrations of ownership below 10 percent and 20 percent and the small number of shareholdings in excess of 20 percent of companies on the NYSE.

The plots suggest that regulation has a significant influence on control patterns in different countries. Shareholder protection, through measures that curb voting power or make it hazardous to exercise, and anti-takeover rules have given rise to a management control bias in the United Kingdom and United States. Weak minority protection and leverage control devices have created a private control bias in Germany and, still more significantly, voting blocks are concentrated around critical levels determined by regulatory rules in different countries.

c.   Ownership of Voting Blocks

Not only does the scale of corporate control differ appreciably across countries but so too do the parties who exert it. Table 2 shows the number of blocks and

Table 2: Voting Blocks by Blockholder Type

Range	Austria [1] No.	Min.	Mean	Med.	Max.	Germany [2] No.	Min.	Mean	Med.	Max.	Spain [3] No.	Min.	Mean	Med.	Max.
Government	9	24.0	53.1	51.0	81.6	18	8.2	45.3	40.7	99.0	37	5.7	46.8	49.0	95.2
Banks	11	6.4	42.0	41.9	100	77	5.1	23.8	15.0	99.0	48	5.0	21.2	13.6	91.5
Insurance						34	5.0	11.9	20.1	96.7	56	5.0	20.8	14.6	91.5
Families/individ.	45	5.0	26.0	12.3	100	205	5.0	26.9	18.2	100	163	5.0	16.0	9.5	87.5
Domestic firms	10	6.6	39.4	51.5	64.3	180	5.0	61.6	70.6	100	203	5.0	24.1	16.7	98.0
Foreign firms	26	5.7	31.6	18.7	87.0						125		20.7	9.1	97.2
Assoc./pools						21	5.9	45.2	49.1	100					
Holding						53	6.9	52.9	50.3	100					
Investment firm						36	5.5	25.1	40.0	99.0					
Bank rel. inv. firm						5	10.2	18.1	11.0	41.4					
Foundation						16	8.0	50.1	51.6	98.1					
Other						3	13.0	18.9	20.2	23.6					
All blocks	101	5.0	33.1	22.7	100	648					632	5.0	20.7	12.3	98.0

Range	Italy [4] No.	Min.	Mean	Med.	Max.	The Netherlands [5] No.	Min.	Mean	Med.	Max.	United Kingdom [6] No.	Min.	Mean	Med.	Max.
Government	34	0	6.8	0	97.4	48	0.0	4.4	0.0	39.8	6		6.7		
Banks	156	0	9.5	0	95.6	34	0.0	8.3	0.0	93.0	71		5.1		
Insurance	13	0	1.1	0	93.9	36	0.0	8.9	0.0	97.1	226		4.0		
Families/individ.	234	0	20.1	0	95.4	22	0.0	1.4	0.0	27.0	61		5.2		
Domestic firms	160	0	20.3	2.0	100						102		10.6		
Foreign firms	116	0	9.1	0	99.9										
Invest./pen. fund	57	0	0.8	0	8.9	6	0.0	0.4	0.0	19.0	474		7.0		
Exec. directors											117		4.5		
Non-exec. direc.											184		5.0		
Real estate											1		0.1		
Other financ. inst.	18	0	1.1	0	66.9	61	0.0	11.1	0.0	85.6					
State						4	0.0	1.1	0.0	50.0					
Admin. office						54	0.0	26.9	0.0	100					
Total	788	0	68.4	71.5	100										

Source: [1] Gugler et al. (2001: Table 7, data for 1996); [2] Becht und Böhmer (2001: Table 5, data for 1996); [3] Crespí-Cladera and García-Cestona (2001, data for 1995); [4] Bianchi et al. (2001, data for 1996); [5] De Jong et al. (2001: Table 6, data for 1996); [6] Goergen and Renneboog (2001, data for 1992).

the mean, median, minimum, and maximum size of blocks held by the different classes of investors in Austria, Germany, Italy, Spain, the Netherlands, and the United Kingdom. Figures 5 and 6 show the number of reported blocks owned by different classes of investors in U.K., German, and Austrian companies.

The United Kingdom: As is well known, financial institutions, pension funds, and life insurance companies are the dominant class of shareholders in the United Kingdom. Figure 5 records that they hold 62 percent of the recorded blocks in the United Kingdom. While financial institutions dominate in terms of numbers of blocks, Table 2 records that the size of blocks held is relatively small. The median size of blocks held by insurance companies is 4.0 percent and by investment and pension funds is 7.0 percent.

Austria and Germany: Figure 6 shows that in Germany and Austria, families and individuals and other companies have the largest blockholdings. In Germany, individuals and families hold 32 percent of blocks, other companies 28 percent, trusts and holding companies 21 percent, financial institutions 17 percent, and government 3 percent. In Austria, individuals and families hold 45 percent of blocks, companies 36 percent, financial institutions 11 percent, and government

Figure 5: Percentage of Voting Blocks Associated with Different Types of Investors in the United Kingdom[a]

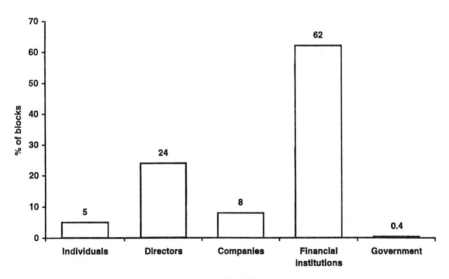

[a]Financial institutions, pension funds, and life insurance companies are the dominant class of shareholders in the ultimate blocks. They hold 62 percent of the recorded blocks in the United Kingdom. Directors hold 24 percent of the blocks, companies 8 percent, and individuals 5 percent. For the raw counts, see Table 2.

Figure 6: Percentage of Voting Blocks Associated with Different Types of Investors in Germany and Austria[a]

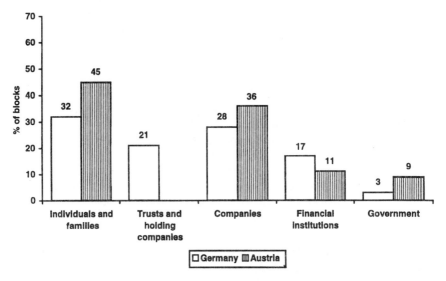

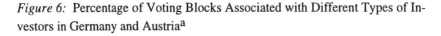

[a]In Germany and Austria, families and individuals and other companies dominate the blockholdings. In Germany, individuals and families hold 32 percent of blocks, other companies 28 percent, trusts and holding companies 21 percent, financial institutions 17 percent, and government 3 percent. In Austria, individuals and families hold 45 percent of blocks, companies 36 percent, financial institutions 11 percent, and government 9 percent. For the raw counts, see Table 2.

9 percent. The median size of block held by families is 26.9 percent in Germany and 26.0 percent in Austria. The median size of blocks held by companies is much larger than that of families in Germany (61.6 percent) and somewhat larger in Austria (39.4 percent for domestic firms).

Italy: Table 2 records a dominant role for families and domestic firms in Italy. Pyramidal holdings are widespread and are primarily associated with holdings by families, coalitions of corporate shareholders, and the state. Financial institutions, including banks, have only played a limited role in the ownership of Italian companies.

The Netherlands: In the Netherlands, there are a substantial number of large blocks held by administration offices that are often controlled by the boards of the companies they control.

Belgium: Foreign ownership is an important feature of Belgian corporate control, in particular from France and Luxembourg. Control is sometimes exerted via pyramid structures and, as in other Continental European countries, a range of anti-takeover devices are employed. The case of Solvac S.A. illustrates these (see Figure 7). Solvac S.A. is a listed company but has registered shares that can only be held by private investors. Solvac S.A. has entered into an agreement with Sofina S.A., Deutsche Bank AG, and Générale de Banque S.A. to ward off any hostile takeover bids for Solvay S.A. Sofina S.A. is controlled by the Boel, Solvay, and Janssen families.

Figure 7: Majority Control and Concentrated Ownership: Solvay S.A.[a]

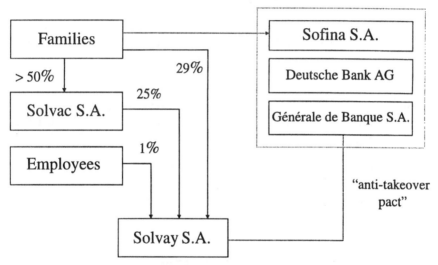

[a]Solvay S.A. has three known shareholder groups: the Boel, Solvay, and Janssen families (29 percent), Solvac S.A. (25 percent), and the company's employees (1 percent). Solvac S.A. is a holding company that has been formed to control Solvay S.A. Although Solvac is listed, it has issued registered shares that can only be held by private investors. Solvac's stock is illiquid and on 31 December 1998 it was estimated to trade, compared to the company's intrinsic value, at a discount of 28.6 percent (Banque Degroof 1999). Solvac S.A. introduces a small degree of pyramiding into this control structure. Through Solvac S.A., the families exert majority control. To protect the company further, for example, against family disputes, Solvay has entered an anti-takeover pact with "friends" (Sofina S.A., Deutsche Bank AG, and Générale de Banque S.A.). The Boel, Solvay, and Janssen families control Sofina S.A. (Banque Degroof 1999).

Spain: In Spain, families and firms are again the largest holders of share blocks. The average size of the blocks is similar to that of Italian companies (a median of 16.0 percent for family holdings and 24.1 percent for domestic firms). The largest holdings have traditionally been associated with the state.

In sum, financial institutions are therefore the largest holders of voting blocks in the United Kingdom but their blocks are small on average. In contrast, in most Continental European companies there are substantial voting blocks in the hands of families and companies, and there are a variety of mechanisms intensifying this through the separation of cash flow and voting rights.

However, there is another feature of corporate control in many European countries and that is the ability of management to entrench themselves. In some companies, there are no identifiable owners or owners are disenfranchised through intermediary institutions or lock-in devices. In Austria there are a significant number of companies with no identified owner. For example, the holder of the largest block of more than 40 percent of Bank Austria is an ownerless association, Anteilsverwaltung Zentralsparkasse (AVZ). In the Netherlands, the largest blocks of shares are held by "administrative offices." These issue depository shares that give certificate holders the right to attend and speak at shareholders' meetings and to call for extraordinary meetings. But they have no votes; voting rights attached to shares can only be exercised by administrative offices. Often the boards appoint themselves (the so-called "structural regime" that is compulsory for companies with certain characteristics). In Spain and France, voting right restrictions provide formidable protection from shareholder influence and control contests.

An alternative protection device recorded in the Netherlands is to issue preferred shares to friendly investors. These shareholders have the right to make a binding nomination for the appointment of management. "Potential capital" is a further device for preventing transfers of control. They are like poison pills, except that issued capital goes to friendly investors, for example, foundations, in the event of a hostile bid.

Unilever illustrates the operation of these protection devices (see Figure 8). Unilever comprises Unilever N.V.—the Dutch part—and Unilever PLC—the U.K. part. They trade as a single entity. This is achieved through two holding companies, N.V. Elma and United Holdings Limited, which are held in turn by the Unilever companies and have cross-shareholdings in each other. They in turn hold special shares and deferred stock in Unilever NV and PLC, respectively. The significance of these special shares and deferred stock is that only they can nominate persons for elections as members of the Boards of Directors of NV and PLC. In other words, elections to the board of Unilever are by two companies fully owned by Unilever. This is said to be required to "ensure unity of management of the Unilever Group" (Unilever's 20F declaration page 33).

There are far fewer takeover defences available to companies in the United Kingdom, but Table 2 reports that there are a large number of share blocks held by both executive and nonexecutive directors. In the presence of highly dispersed

Figure 8: Management Control through Legal Devices: Unilever Plc/NV[a]

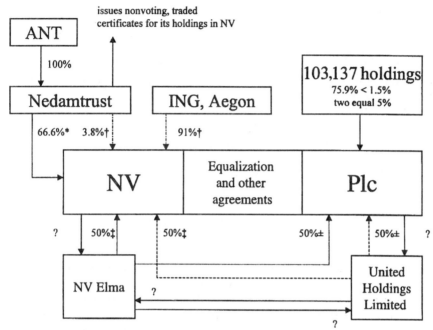

[a]Unilever Plc/NV has a complicated control structure. NV has locked control into Plc, and vice versa. In addition, NV leverages voting power through a trust company and uses preference shares to fend off unsolicited bids for control. *Cross-lock in*: NV has the sole right to nominate directors for election at Plc, and vice versa. This is achieved through a cross-holding structure involving two holding companies, N.V. Elma and United Holdings Limited, which hold 50% of special shares and deferred shares in NV and Plc. Since N.V. Elma is a subsidiary of NV and United Holdings Ltd of Plc, Elma cannot vote its special shares of NV and United Holdings cannot vote its deferred shares in Plc (Unilever Form 20-F 1998, page 33). Hence, the other shareholders of NV and Plc can refuse to elect the directors Plc and NV nominate, but they cannot nominate other directors for election. In addition, "each of Unilever's ten directors is a full-time executive and is a director of both NV and Plc" (Unilever Annual Review 1998, English Version). *Voting power leverage*: NV has issued 3 classes of shares; special shares (‡), ordinary shares (*), and three types of cumulative preference shares (7%, 6%, 4%; †), representing 0.3%, 70.5%, and 29.2% of par value, respectively. Nedamtrust holds 66.6% of the ordinary shares and 3.8% of the cumulative preference shares. For its 66.6% stake in the ordinary shares it has issued nonvoting certificates that are traded on the stock market. Nedamtrust itself is owned and controlled by ANT. *Veto power of preference shareholders*: ING and Aegon hold approx. 91% of the cumulative preference shares. In the unlikely event that a bidder would acquire the ordinary shares of Plc or NV, the bidder would be unable to break up the "equalization and other agreements" that bind Plc and NV without the consent of 2/3 of the preference shareholders.

shareholdings these share blocks may provide management with a significant degree of protection.

There is therefore an alternative to the above conclusions of private control bias on the Continent and management control bias in the United Kingdom and United States. There is a high degree of management entrenchment in all countries, either through lock-in vehicles on the Continent or through antitakeover mechanisms in the United Kingdom and United States. Market control or control by a blockholder who has the best interest of all shareholders in mind are rare.

d.   Changing Patterns of Control

This paper provides a snapshot of the voting control of European companies at the end of the 20th century. But there is evidence that these patterns are changing. The Agnelli group, for example, has disposed of companies operating in chemicals and cement. Banks are playing a larger role, particularly in companies in financial distress. Takeover legislation requiring mandatory bids when shareholdings reach 30 percent is encouraging concentrations of shareownership around this level, as has already been documented for the United Kingdom. In Spain, the significance of the state has declined with privatizations. Some state control has been maintained through golden shares but these have not been widely applied. In Sweden, cross-shareholdings have largely disappeared. Dual class shares have been eliminated in many firms. There has been more intervention by minority shareholders in, for example, blocking the merger between Volvo and Renault.

With the dismantling of mechanisms for separating cash flow from voting rights comes a move away from private control biases and neutral control. It is unclear how far these processes will go. It is even more unclear whether they will give rise to market control with freely operating markets in corporate control or simply tilt Continental Europe to a management control bias with U.S. style managerial entrenchment through anti-takeover devices. The evidence of a free market in corporate control emerging in Continental Europe is limited to date. There is much antipathy towards markets in corporate control both on social grounds and for the short-termist reasons discussed above in relation to "influence" and investment "realization" periods. If control leverage devices are extinguished and voting control as well as cash flow rights become more widely dispersed, we may well see the emergence of more managerial lock-in mechanisms to take their place. It may, for example, be no coincidence that voting caps are reported to be on the rise in France (Iskander and Targett 2000).

4.   Conclusion

Section 2 described three conflicting views of the influence of regulation on ownership and control of corporations. According to Roe (1994), dispersed ownership and control in the United States result from a populist agenda of imposing regulatory impediments on concentration of control in the hands of a small number of investors. According to La Porta et al. (1999), the dominance of large shareholder blocks in most countries in the world is a reflection of inadequate investor protection. According to Easterbrook and Fischel (1991), regulation reflects the needs of managers and investors. Roe's thesis implies that corporate governance deficiencies result from overregulation, La Porta et al.'s implies that improved corporate governance will come from more regulation, while Easterbrook and Fischel's suggests that Europe and the United States have the regulation and corporate governance systems that are best suited to their activities.

What emerges from this paper is a remarkable contrast with the traditional, Anglo-American view of the corporation. Concentration of voting control is strikingly higher in Continental European companies than in their U.K. or U.S. counterparts. Control is concentrated not only because of the presence of large investors or core investor groups, but also because of the absence of significant holdings by others. In contrast, in the United Kingdom and United States, not only are there few large shareholders but also the second, third, and smaller shareholdings are not appreciably smaller than the largest. This gives rise to the possibility of effective control through coalitions, but not by individual shareholders. However, since a typical block in the United Kingdom is twice as large as in the United States (10 percent versus 5 percent), the potential for coalition control is larger in the United Kingdom than in the United States.

Still more striking than differences in average sizes of shareblocks is the complete distribution of the largest shareholdings. In most Continental European countries, there is a fairly uniform distribution of the largest voting blocks. In contrast, in the United Kingdom and United States there is a strong "market/ management bias" towards dispersed control. However, it would be wrong merely to contrast Continental European with Anglo-American control. There is a marked variation within Europe, ranging from a "private control bias" in Germany to a modest management control bias of the Anglo-American variety in the Netherlands and Spain. Indeed, the largest Spanish companies already combine complete protection of management from takeovers with a very broad shareholder base, just like many of their U.S. counterparts. Still more interesting is the concentration of voting blocks around certain critical levels—blocking minority and super-majority holdings in Austria and Germany and majority control in several other countries. Regulation has affected the entire pattern and distribution of corporate control in all countries. As a general proposition, we believe that control is concentrated in forms in which regulation confers particular advantages: shareblocks are concentrated at levels at which there are significant control benefits.

We suggested in Section 2 that this may be important in terms of the relation between the control and the activities of firms. In principle, market control efficiently reallocates control to those who derive the greatest benefit from exerting it. However, it also limits the period for which anyone can expect to be able to retain control ("influence periods"). Market control therefore efficiently reallocates control of projects with short "realization periods" but may discourage the implementation of projects with long realization periods. According to this view, the relevance of corporate control for real activities is a reflection of the relation of influence to realization periods. Different types of corporate control and therefore regulatory arrangements are suited to different forms of corporate activities: industries whose investments have short realization periods thrive in systems with market control, whereas those with long realization periods benefit from management control.

One of the troublesome, or puzzling, implications of our analysis is the fragility of market control. Even in countries where voting power and ownership are dispersed, management control is often not contestable. Devices such as poison pills, anti-takeover charter provisions, options for issuing voting stock to friendly parties, and voting caps can be used to limit the control that external investors can exert. Voting power dispersion overestimates the importance of external control in countries, such as the United Kingdom and United States, which may have a management, not a market control bias. The primary distinction is not then between market and private control, but between management and private control. Do projects with short realization periods not find enough financing or is market control rare because such projects are rare?

Even in countries with dominant shareholders, we have observed a variety of mechanisms that management can employ to protect itself from external investor interference. If, as a consequence, corporations are run by managers who are able to shield themselves from external influence by investors, then control by owners will be largely irrelevant. The Berle and Means view of strong managers/weak owners may therefore be applicable even in the presence of dominant blockholders and the formal distinction between patterns of corporate control may be of little relevance in the face of a class of largely unaccountable management. Whether this alternative description is correct is yet to be established, but regardless of whether it is, this paper serves as a reminder of the multifarious nature of capitalism within, let alone outside, Europe.

Bibliography

Agnblad, J., E. Berglöf, P. Högfeldt, and J. Svancar (2001). Ownership and Control in Sweden: Strong Controlling Owners, Weak Minorities, Sound Control. In F. Barca and M. Becht (eds.), *The Control of Corporate Europe*. Forthcoming. Oxford: Oxford University Press.

Aoki, M. (1990). Toward an Economic Model of the Japanese Firm. *Journal of Economic Literature* 28 (1): 1–27.

Barca, F., and M. Becht (2001). *The Control of Corporate Europe*. Forthcoming. Oxford: Oxford University Press.

Barca, F., M. Bianchi, F. Brioschi, L. Buzzacchi, P. Casavola, L. Filippa, and M. Paganini (1994). *I Modelli di Controllo e La Concentrazione Proprietaria Messi a Confronto Empiricamente.* III vols. Vol. II, *Assetti, Proprietà e Controllo Nelle Imprese Italiane Medio-Grandi.* Bologna: il Mulino.

Bebchuk, L. (1999). A Rent-Protection Theory of Corporate Ownership and Control. NBER Working Paper 7203, Cambridge, Mass.

Bebchuk, L., and M. Roe (1999). A Theory of Path Dependence in Corporate Ownership and Governance. *Stanford Law Review* 52(1): 127–170.

Becht, M. (2001). Beneficial Ownership in the United States. In F. Barca and M. Becht (eds.), *The Control of Corporate Europe*. Forthcoming. Oxford: Oxford University Press.

Becht, M., and E. Böhmer (2001). Ownership and Voting Power in Germany. In F. Barca and M. Becht (eds.), *The Control of Corporate Europe*. Forthcoming. Oxford: Oxford University Press.

Becht, M., A. Chapelle, and L. Renneboog (2001). Shareholding Cascades : The Separation of Ownership and Control. In F. Barca and M. Becht (eds.), *The Control of Corporate Europe*. Forthcoming. Oxford: Oxford University Press.

Berle, A.A. (1958). "Control" in Corporate Law. *Columbia Law Review* 58: 1212–1225.

Berle, A.A., and G.C. Means (1932). *The Modern Corporation and Private Property.* New York: The MacMillan Company.

Bianchi, M., M. Bianco, and L. Enriques (2001). Pyramidal Groups and the Separation Between Ownership and Control in Italy. In F. Barca and M. Becht (eds.), *The Control of Corporate Europe*. Forthcoming. Oxford: Oxford University Press.

Black, B.S. (1990). Shareholder Passivity Reexamined. *Michigan Law Review* 89: 520.

Bloch, L., and B. Kremp (2001). Ownership and Voting Power in France. In F. Barca and M. Becht (eds.), *The Control of Corporate Europe*. Forthcoming. Oxford: Oxford University Press.

Carlin, W., and C. Mayer (2000a). Finance, Investment and Growth. Mimeo, University of Oxford.

Carlin, W., and C. Mayer (2000b). How Do Financial Systems Affect Economic Performance. In X. Vives (ed.), *Corporate Governance: Theoretical and Empirical Perspectives.* Cambridge: Cambridge University Press.

Corbett, J., and T. Jenkinson (1996). The Financing of Industry, 1970–1989: An International Comparison. *Journal of the Japanese and International Economies* 10(1): 71–96.

Crespi-Cladera, R., and M.A. García-Cestona (2001). Ownership and Control of Spanish Listed Firms. In F. Barca and M. Becht (eds.), *The Control of Corporate Europe.* Forthcoming. Oxford: Oxford University Press.

De Jong, A., R. Kabir, T. Marra, and A. Roell (2001). Ownership and Control in the Netherlands. In F. Barca and M. Becht (eds.), *The Control of Corporate Europe.* Forthcoming. Oxford: Oxford University Press.

Easterbrook, F.H. (1997). International Corporate Differences: Market or Law? *Journal of Applied Corporate Finance* 9(4): 23–29.

Easterbrook, F.H., and D.R. Fischel (1991). *The Economic Structure of Corporate Law.* Cambridge, Mass: Harvard University Press.

Edwards, J., and K. Fischer (1994). *Banks, Finance and Investment in Germany.* Cambridge: Cambridge University Press.

Franks, J., and C. Mayer (1995). Ownership and Control. In H. Siebert (ed.), *Trends in Business Organization: Do Participation and Cooperation Increase Competitiveness?* Tübingen: Mohr Siebeck.

Goergen, M., and L. Renneboog (2001). Strong Managers and Passive Institutional Settings in the UK. In F. Barca and M. Becht (eds.), *The Control of Corporate Europe.* Forthcoming. Oxford: Oxford University Press.

Grossman, S., and O. Hart (1980). Takeover Bids, the Free-Rider Problem and the Theory of the Corporation. *Bell Journal of Economics* 11(1): 42–64.

Gugler, K., S. Kalss, A. Stomper, and J. Zechner (2001). The Separation of Ownership and Control : An Austrian Perspective. In F. Barca and M. Becht (eds.), *The Control of Corporate Europe.* Forthcoming. Oxford: Oxford University Press.

Hoshi, T., A. Kashyap, and D. Scharfstein (1991). Corporate Structure, Liquidity, and Investment: Evidence from Japanese Industrial Groups. *Quarterly Journal of Economics* 106(1): 33–60.

Iskander, S., and S. Targett (2000). French Groups Curb Voter Rights. *Financial Times,* April 27.

Jensen, M.C. (1989). The Eclipse of the Public Corporation. *Harvard Business Review* 67: 61–74.

Jensen, M.C., and W.H. Meckling (1976). Theory of the Firm: Managerial Behavior, Agency Costs and Ownership Structure. *Journal of Financial Economics* 3(4): 305–360.

Kang, J.-K., and R. Stulz (2000). Do Banking Shocks Affect Firm Performance? An Analysis of the Japanese Experience. *Journal of Business* 73: 1–23.

Kaplan, S.N., and B.A. Minton (1994). Appointments of Outsiders to Japanese Boards: Determinants and Implications for Managers. *Journal of Financial Economics* 36(2): 225–258.

La Porta, R., F. Lopez-de-Silanes, A. Shleifer, and R.W. Vishny (1997). Legal Determinants of External Finance. *Journal of Finance* 52(3): 1131–1150.

La Porta, R., F. Lopez-de-Silanes, and A. Shleifer (1999). Corporate Ownership around the World. *Journal of Finance* 54(2): 471–517.

La Porta, R., F. Lopez-de-Silanes, A. Shleifer, and R.W. Vishny (2000). Investor Protection and Corporate Governance. *Journal of Financial Economics* 58(1/2): 3–27.

Manne, H. (1965). Mergers and the Market for Corporate Control. *Journal of Political Economy* 23(2): 110–120.

Mayer, C. (1988). New Issues in Corporate Finance. *European Economic Review* 32(5): 1167–1183.

Mayer, C. (2000). Ownership Matters. Brussels. Inaugural Lecture, Oxford University.

Prowse, S.D. (1992). The Structure of Corporate Ownership in Japan. *Journal of Finance* 47(3): 1121–1140.

Roe, M. (1991). A Political Theory of American Corporate Finance. *Columbia Law Review* 91(1).

Weinstein, D.E., and Y. Yafeh (1998). On the Costs of a Bank-Centered Financial System: Evidence from the Changing Main Bank Relations in Japan. *Journal of Finance* 53(2): 635–672.

III.

New Conditions for
Macroeconomics and Growth

Paul De Grauwe and Magdalena Polan

F31   E31

(selected Countries)   F32 F36

Increased Capital Mobility: A Challenge for National Macroeconomic Policies

E62

E52

1.   Introduction: How Important Is the Increase in Capital Mobility?

Influenced by the hype about globalization, many observers take it for granted that the degree of capital mobility has increased substantially during the last decades and that capital markets are more integrated today compared to any previous period in history. The empirical evidence, however, is not as clear-cut. In this section, we briefly survey this evidence.

There is a very large literature following up on the seminal paper of Feldstein and Horioka (1980). Feldstein and Horioka reasoned that, in a world of perfectly mobile capital, domestic savings would seek out the highest returns in the world capital market and would be independent of local investment demand. By the same mechanism, the world capital market should serve as a source of financing for domestic investment needs. Thus, if capital markets are integrated, the investment ratio (investments/GDP) should be independent of the savings ratio. Feldstein and Horioka argued that the correlation between investment and saving in a cross-section of countries might provide a test of global (or international) capital mobility. They found that this correlation was very high in the postwar period, indicating that the degree of capital mobility was not substantial.

Subsequent econometric research has confirmed the high correlation between savings and investment ratios in cross-country regressions. However, it has also been stressed by several researchers that a high correlation between savings and investment ratios may not necessarily signal a low degree of financial integration (see Obstfeld 1986, Summers 1988, Barro et al. 1992, Frankel 1991). Two alternative explanations have been advanced. The first one is that common factors such as population growth, output changes, or productivity shocks may determine

---

Remark: We are grateful to Jean Pisani-Ferry and to the participants of the conference for their helpful comments.

investment and savings simultaneously. When these common shocks occur the two variables are cointegrated and automatically exhibit a high correlation. This will be the case even if capital markets are fully integrated. The second explanation relies on the fact that economic policies tend to be similar across countries. Policymakers in most countries may seek to attain approximate current account balance. This goal can be achieved through appropriate monetary and fiscal policies. If cross-country targets are similar, then the high correlation of savings and investments across countries follows automatically.

One conclusion from the vast literature is that the original Feldstein–Horioka test is not informative, since conventional cross-sectional regressions are likely to produce high saving-investment correlations, regardless of whether the degree of international capital mobility is high or low (see Jansen 1996).

In addition to its methodological and econometric problems, the Feldstein–Horioka econometric test does not produce a benchmark that can indicate low or high integration. Even if the Feldstein–Horioka criterion measures integration properly and the econometrics yields a proper estimate, we are still left without a yardstick telling us what is "high" and what is "low." However, potentially useful information can be obtained by analyzing the changes over time in the correlation between savings and investment. Such an analysis has been done by Taylor (1994, 1996). Using a modified F–H measure, Taylor found a lower correlation between high-income countries. In addition, his analysis uncovered a general decline in the correlation (higher integration of markets) from 1980 onwards. Taylor concludes that in this modified framework, international markets do exhibit a recent tendency towards increased integration. Although this conclusion seems plausible, it is subject to a similar criticism as the one leveled against the original Feldstein–Horioka result. The decline in the correlation observed since 1980 could be due to a decline in the importance of common shocks.

Recently economists have taken a long historical view to analyze the question of whether capital mobility has increased. A surprising finding is that net capital flows (as measured by the current account) tended to be of the same order of magnitude during the period of the international gold standard as now. This has been confirmed by Zevin (1992), Sachs and Warner (1995), and Rodrik (1998), leading to the conclusion that today's degree of capital market openness is nothing particular as compared to the situation a century ago. Using U.S. data, Eichengreen (1999) has claimed, however, that the present degree of financial integration has increased relative to one hundred years ago.

To sum up, the degree of financial market integration today is not dramatically higher than one hundred years ago. This runs counter to the conventional wisdom that exists in the popular press and in the large "globalization literature." Nevertheless, there seems to be evidence that the degree of financial integration in the world has been increasing in the last few decades. The recent increase leads to many new challenges, one of which has to do with the conduct of national monetary and fiscal policies.

2.   Implications for Monetary and Fiscal Policies

There is a general consensus among economists that the increase in capital mobility has affected the viability of fixed exchange rate regimes. In particular, it has made the fixed exchange rate regime more fragile and less capable of withstanding speculative movements. The very fact that speculators expect a future devaluation can dramatically increase the cost of defending a fixed exchange rate, giving strong incentives to the monetary authorities not to fight the speculators and to devalue. This phenomenon has been given theoretical backing in the so-called second generation models of speculative attacks (see Obstfeld 1996). In addition, increased capital mobility may have intensified the contagious effects of crises (see Eichengreen and Wyplosz 1993). As a result, the increasing integration of financial markets has forced more and more countries to move away from fixed exchange rate arrangements. This phenomenon is well illustrated in Table 1. Since 1975, the number of developing countries pegging their currencies has dropped by half.

Table 1: Developing Economies: Evolution of Pegged Exchange Rate Arrangements

Year	Percentage of developing countries pegging their currencies	Of which pegging to (in percent)				
		USD	FF	SDR	Other currency	Basket
1975	88	50	15	12.5	12.5	10
1986	70	37	27	16	13	7
1998	43	34	26	21	14	5

Source: Mussa et al. (2000).

The implications of the move away from pegged exchange rates are that countries have sought alternative arrangements in drastically different directions. One set of countries (the largest part) has sought refuge in more flexible exchange rate regimes. This is shown in Table 2.

Another group of countries has chosen much tighter arrangements than pegged exchange rates. The most notable move was made by eleven EU countries that decided to abolish their intra-exchange rates altogether and to move into a mone-

Table 2: Exchange Rate Arrangements in Small Economies in 1998

Exchange rate arrangements in 1998	Number of countries	Average GDP of the economy (bill. USD)	Average trade share	Average share of the largest export partner	Fraction of countries with controls on current account
			percent		
Pegged to:	45	1.58	51.8	33.6	78
USD	16	1.20	61.1	29.5	69
FF	13	2.03	34.4	36.9	100
Other	8	1.52	63.4	37.2	75
Basket	8	1.68	53.4	34.1	63
Flexible:	28	2.15	51.3	34.3	57
Managed float	11	2.00	69.7	27.7	64
Independent float	17	2.25	38.7	38.9	53
Small economies	73	1.80	51.6	33.9	70

Source: Mussa et al. (2000).

tary union on January 1, 1999. Other countries sought to tighten up their exchange rate arrangements by adopting currency boards.

The consensus today is that countries have little choice but to move to one of the two extremes, i.e., either towards more flexibility or towards more rigidity, because the intermediate regime of pegging is not sustainable for long in a world of high capital mobility.

These new policy choices create new challenges for macroeconomic policies that we analyze in the next sections.

3.   The Challenge of Increased Flexibility of the Exchange Rates

The major challenge here is how countries can anchor nominal variables such as the price level and the money stock, once the exchange rate anchor has gone. There is no doubt that in the past many countries used the fixed exchange rate as the anchor for their domestic price level and money stock. In doing so, they in fact used the services of the leader in the system (for most countries this was the

United States), who was doing the job of explicitly anchoring nominal variables. In a flexible exchange rate environment, these countries cannot rely anymore on the economic policy of another country and have to manage the price level and the money supply themselves.

The need to anchor nominal variables in a flexible exchange rate environment explains the increasing popularity of procedures of explicit inflation targeting. We now observe that more and more countries have switched towards inflation targeting. Among the industrial countries, seven have adopted inflation targeting during the past decade. They are (in chronological order): New Zealand, Canada, the United Kingdom, Finland, Sweden, Australia, and Spain. With the introduction of the euro on January 1, 1999, the central banks of Finland and Spain lost their power to conduct independent monetary policy and transferred it to the European Central Bank. All of the listed countries are small or middle-sized open, industrialized economies. They have shown a poor record in fighting inflation (for industrialized countries standards) over the past 30 years and they were generally perceived to lack monetary policy credibility.

In all these countries, the inflation target is set around 2 percent. In Australia (and Finland before 1999), central banks target the point objective, while in Canada, New Zealand, the United Kingdom, and Sweden central banks specify a range for the inflation target. The Spanish central bank used to specify a ceiling for the inflation rate.

The empirical evidence indicates that inflation targeters have had some success in anchoring their domestic price levels. Many issues remain, however. First, other industrial countries without explicit inflation targeting procedures have been equally successful in reducing inflation. It is, therefore, unclear whether inflation targeting is superior to other forms of monetary control. Second, there is the question for inflation targeters and others alike of whether the success in reducing inflation is sustainable. A subsidiary question is whether the success in stabilizing the price level will also lead to more stability of the exchange rates. Since price levels are one of the fundamental variables determining exchange rates, one would expect that success in stabilizing price levels would also tend to stabilize exchange rates.

The evidence on purchasing power parity, however, is weak. This is especially the case for currencies experiencing low inflation rates. The recent developments in the euro-dollar exchange rate illustrate this point. Inflation differentials between the United States and Euroland have remained extremely low since the start of EMU. They fluctuated around 1 percent to 1.5 percent (see Figure 1), while the U.S. dollar appreciated by more than 20 percent during the same period. Surprisingly, it is the United States with the appreciating currency, which experienced a slightly higher rate of inflation. One can conclude that it is far from certain that price stability will do much to stabilize exchange rates. We are likely

Figure 1: Inflation in the United States and Euroland

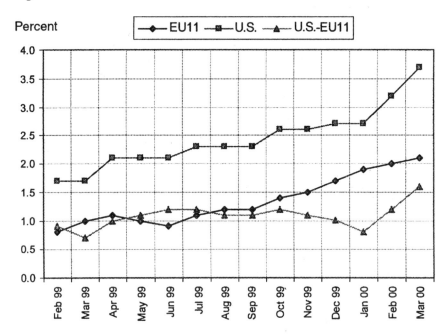

to have to live with significant volatility of the major exchange rates even in a world of price stability.

This leads to the other challenge for the countries moving towards more exchange rate flexibility. The first question that arises in this context is whether countries should worry about the kind of exchange rate volatility that such a system implies. The second question is whether they can and should do something about this.

To the first question, the answer seems to be yes, but only when misalignments take on large proportions. Examples of such large misalignments have been the dollar movements of 1980–1985, and the recent movements of the euro-dollar and euro-yen rates. It should be stressed that these large misalignments have been relatively rare.

The next question, which has been, and continues to be hotly debated, is: what can be done about the large swings in the exchange rates? One school of thought (represented by Williamson, Bergsten and others) has been arguing that agreements to set up looser forms of exchange rate pegs (target zones) are called for. In other words, these are proposals aiming at escaping the iron logic of a world of capital mobility that drives countries to the extremes of high flexibility or rigid fixity.

It is doubtful that these target zone agreements can survive any better than pegged exchange rate systems. Although target zones provide more flexibility than pegged rate systems, they suffer from the same weakness, i.e., the national monetary authorities must be willing to subordinate domestic objectives to the external constraint of the target zone. To take an example, suppose that today, in June 2000, a target zone agreement existed between the United States and Euroland. This would be fine for Euroland because the euro depreciation would act as a signal for the ECB to tighten monetary policy, which it probably wants to do anyway because of the inflationary pressures generated by the weak euro. But what about the United States? There can be no doubt that a target zone agreement today would lead to a major policy conflict for the U.S. monetary authorities. The reason is that it would force the U.S. monetary authorities to loosen monetary policies to bring down the dollar. Such an easy monetary policy, however, would further exacerbate the domestic consumption boom and increase inflationary pressures. Faced with such a dilemma, there is no doubt that the Fed would opt for its domestic objectives and would not do its part of the deal implicit in a target zone agreement. Such conflicts between domestic objectives and external constraints would emerge regularly, reducing the credibility of the target zone agreement.

One can conclude that the only cooperative initiatives that have some chance of success will be ad hoc agreements concluded when there is a broad consensus that exchange rate movements "have gone too far," and that it is in the interest of all concerned sides to correct these movements. This happened in 1985, and led to the Plaza agreement, which was relatively successful in correcting the overvaluation of the dollar. Such ad hoc agreements are all one can wish for in the future.

4.   The Challenge of the Fixers

Many countries have decided that the optimal response to greater capital mobility is not to go for more flexibility, but to look for tighter fixity in the exchange rate. European countries forming EMU went all the way with this logic and started a monetary union. In Section 7, we deal with some of the challenges that these countries face. In this section, we discuss the experiments of other countries that chose to go in the direction of more fixity.

A significant number of countries have decided to take up a monetary regime—the currency board—that seemed to be discredited, since it was linked to colonial times. Surprisingly, countries that have chosen this monetary regime seem to have performed relatively well compared to other countries choosing a different fixed exchange rate regime. This has been documented in a recent IMF study (see Ghosh et al. 1998). The study compared inflation and output in countries operating under different exchange rate arrangements (a currency board, a fixed exchange rate regime, and a flexible exchange rate regime). Ten countries were included in the sample of countries operating under a currency board: Antigua and Barbuda, Argentina, Djibouti, Dominica, Estonia, Grenada, St. Lucia, St. Vincent and the Grenadines, Hong Kong, and Lithuania. Nine countries in the sample linked their currencies to the U.S. dollar and one to the euro (Estonia). Other currency board countries were not included in the sample because of the lack of data or very short time of their operation. The study covered the period from 1970 to 1996.

Inflation averaged 6 percent per year in countries with a currency board, which was substantially lower than in countries with a flexible exchange rate (50 percent) and other forms of a peg (20 percent). The results remain unchanged after adjustment for outliers.

Controlling for other variables like real income, the growth rate of the money stock, the rate of turnover of the central bank governor and the ratio of the sum of exports and imports to GDP, the estimated inflation differential remained highly significant. The IMF study estimated that a currency board system creates a "confidence effect," amounting to a 3.4 percentage point decrease in inflation rate and caused by the very existence of a currency board in a country.

This favorable inflation performance does not seem to have been bought by lower economic growth. In the sample of countries operating under a currency board, the average annual growth of per capita income was almost twice as high as in all other countries. In addition, the volatility of the GDP growth was slightly lower in currency board countries than in other countries.

These are impressive achievements. Yet, the recent financial crises illustrate that the currency board system does not shield countries from speculative attacks, which in a number of cases (e.g., Hong Kong) have been quite fierce. Although these countries were able to resist the onslaught of the last crises, it is unclear whether they will be able to do so indefinitely. Thus, a currency board remains a fragile construction. This fragility will be enhanced when the political pressure to take on a monetary regime that makes the country less dependent on a foreign big power accumulates. This pressure is likely to increase as countries regain monetary stability. The big challenge therefore will be to manage the transition to a new exchange rate regime.

5.   The Temptation of Capital Controls

We argued above that the increasing degree of capital mobility puts countries into the uncomfortable situation of having to choose for more flexibility or much tighter arrangements. The temptation to avoid such a choice and to find something in the middle ground will therefore continue to exist. This will also keep capital controls on the political agenda.

One of the more surprising developments is the strong popularity that the Tobin tax has acquired in the political arena. Many parliaments are now voting on motions to introduce such a tax on a worldwide scale. Most NGOs have made of the Tobin tax one of their favorite battle cries against the wicked international speculators.

Will the Tobin tax allow countries to find a middle ground between the two extreme monetary regimes that international capital market integration now increasingly imposes on them? This question can be divided into two subsidiary questions:

- Is the Tobin tax an instrument to reduce exchange rate volatility for those countries that moved in the direction of more flexibility?
- Is the Tobin tax an instrument that can make a pegged exchange rate system less fragile?

a.   The Tobin Tax and Exchange Rate Volatility

Proponents of the Tobin tax have been mesmerized by the large size of the daily transactions in the foreign exchange markets. The latest BIS estimates are that these could amount to $1.5 trillion. Since exchange transactions arising from exports and imports are small, probably not more than 5 percent of the turnover, the quick conclusion has been that all the rest is the result of speculative capital flows. The problem with this conclusion is that it fails to account for an important institutional feature of the foreign exchange market: this is a multidealer market in which the largest part of the daily transactions are done for purposes of hedging and not for purposes of speculation. It has been estimated that 80 percent of the daily flows represent dealers' hedging activities ("hot-potato trading") (see Lyons [1999] on this). Thus, the Tobin tax will discourage short-term speculators *and* short-term hedgers. Since the latter are responsible for (by far) the largest part of the market transactions, it is not obvious that the reduction of hedging activity will tend to reduce the exchange rate variability.

We have to look into the microstructure of the foreign exchange market to give an answer to the question of how the reduction in the size of hedging activities will affect the variability of the exchange rates. The microstructure can be described as follows (for a formal analysis, see Lyons [1999]).

Suppose an individual speculator buys dollars (sells euro), thereby raising S (the price of dollar in units of euro). If there are no dealers, the speculator must find another trader willing to hold the euro. In order to find a risk-averse trader to hold all these euros, the price of euros will have to drop a lot.

With many dealers, we have a different situation. Let us assume a chain of transactions between dealers. The first dealer obtaining the euros will want to unload them, but not the full amount. Because of the drop in the price of the euro the dealer has an incentive to hold a fraction of these cheap euros. Suppose, he holds 5 percent. He then unloads the other 95 percent to another dealer, who has the same incentive to hold a fraction and to unload the rest. (This is the hot-potato trading.) At the end of the line, all the dealers hold a fraction of the initial net speculative demand. (Note that the chain will typically be shorter because a speculator is likely to be found willing to buy the remaining fraction of the euros at the given price.) Since each (risk-averse) dealer holds only a fraction of the initial order flow, he will be willing to accept a smaller price decline of the euro than if any one of them had to hold the full order flow. Thus, when there are many dealers, the price decline of the euro necessary to absorb the initial order flow will be smaller than if there were no dealers.[1]

The previous analysis can also be formulated as follows: The existence of many dealers is a mechanism that allows spreading risk more efficiently. When a speculator buys dollars (sells euros) he forces somebody to take the counterparty risk. When there are many counterparties, this risk can be spread around more efficiently.

If one accepts this reasoning, one comes to the conclusion that taxing all transactions in the foreign exchange market also makes this search for risk spreading more difficult. As a result, it is not certain at all that exchange rates will move less. They could be moving more.

Note that the Tobin tax will discourage hedging in the foreign exchange market more than pure speculation. The reason is that the search for risk spreading involves multiple transactions in the foreign exchange market. For example, assume that the initial sale of euros is 100. Each dealer keeps 5 percent until, say, after five dealers the rest is unloaded to another speculator willing to take a reverse position. The chain of taxes (assuming a Tobin tax of 1 percent) will be:

$$100 \cdot 0.01 \ [1 + 0.95 + (0.95)^2 + (0.95)^3 + (0.95)^4].$$

---

[1]   Note that this argument is based on the concavity of the utility function: risk premia increase with increasing positions in a particular currency.

Thus hedging will be taxed by a multiple of the Tobin tax. In this simple example the hedging activities are taxed at a rate of 4.5 percent. This must have effects on the structure of the market. It is likely to eliminate the multidealer nature of the market and to favor its centralization, like the one that exists in the stock markets. Such a centralization becomes then another substitute for the efficient spreading of risk. It is unclear, however, whether a more centralized market leads to less variability of prices. The evidence seems to go in the other direction, since centralized asset markets experience more price variability (see De Grauwe 1996).

This change in the market structure will also affect another objective of the Tobin tax because it will lead to a large reduction in the size of the daily transactions, as the multiple dealer market tends to disappear. Consequently, the objective to raise revenues for worthwhile international projects will have to be scaled down significantly.

b.   The Tobin Tax as an Instrument to Make Fixed Exchange Rates Less Fragile

Suppose a Tobin tax of 1 percent were imposed on all exchange transactions. (We do not go into the problem of whether such a tax can be implemented in practice. It is quite unlikely that this can be done.) Can such a tax protect a country from a speculative attack of the kind we saw in Asia during 1997–1998? The answer is that it cannot. Speculative attacks on fixed exchange rates are typically driven by expectations of relatively high devaluations. Typical orders of magnitude are 20 percent or more. A tax of 1 percent does very little to discourage these speculative attacks. In addition, much of the capital movements in a crisis situation is the result of panic whereby investors run for the exit door (see Radelet and Sachs 1998). A Tobin tax of 1 percent will do little to discourage panic flows.[2]

One can conclude that the Tobin tax will do little to reduce exchange rate volatility, nor will it give countries that defend a fixed exchange rate a weapon to ward off speculators.[3]

---

[2]   See Tobin (1999), who acknowledges this point.

[3]   There may exist other forms of capital controls that are better suited than the Tobin tax to defend fixed exchange rates. For a discussion, see De Grauwe (2000).

6.   Capital Mobility and Monetary Cooperation

Increased capital mobility induces countries to move in opposite directions in their choice of exchange rate regimes. It, therefore, also leads to opposite requirements as far as policy cooperation in the monetary field is concerned. Those countries that decide to move in the direction of more exchange rate flexibility by the same token reduce the need to coordinate their monetary policies. Countries that move towards tighter exchange rate arrangements find themselves forced to increase monetary cooperation. This is so because in a tightly fixed exchange rate regime (monetary union, currency board) the interest rates of the member countries must be equalized. Some rule must therefore be agreed upon as regards the question of how the joint level of the interest rate will be decided on. In a monetary union, this requires central bankers of the union to sit together and to decide on this jointly. In a currency board regime the country setting up a currency board accepts whatever decision the country to which it ties its currency decides. No such explicit cooperative monetary arrangements are necessary in a flexible exchange rate system.

Thus, the net effect of increased capital mobility on the degree of cooperation of monetary policies in the world is unclear. It all depends on how capital mobility affects the countries' choices of exchange rate regimes. If the net effect of capital mobility is to generalize exchange rate flexibility, then the need for cooperation may actually decline.

While all this is obvious as far as monetary policies are concerned, this is not the case with *fiscal policies*. We turn to this issue in the next section by concentrating on the question of how the move to a monetary union (as in the European Union) affects the need to cooperate in the field of fiscal policies.

7.   The Need for Fiscal Policy Coordination in a Monetary Union

EMU is a quantum jump forward in macroeconomic policy coordination in the EU. Since the start of EMU, decisions about interest rates, money stocks, and reserve requirements have become truly joint decisions. One could not possibly have more cooperation in the monetary field.

The question that arises next is the following. Does the intense cooperation in the monetary field require a parallel intensification of cooperation in the fiscal policy field? In order to analyze this question, let us concentrate on two concepts: spillover effects and asymmetric shocks.

a.   Spillover Effects of Fiscal Policies and the Need to Cooperate

It is well known that gains from cooperation depend critically on spillover effects. If the effects of budgetary shocks in one country on other countries are important, cooperation can improve welfare significantly. When these spillover effects are low, there is little welfare improvement to be expected from budgetary cooperation.

The crucial question therefore becomes the following. How will the existence of EMU affect spillover effects of national fiscal policies? If EMU leads to higher spillovers, then more intense fiscal policy coordination is desirable, and *vice versa*.

EMU can affect spillover effects of fiscal policies in two ways. *First*, EMU could lead to more intense trade links between the member states. The argument that has often been used here is that EMU leads to more price transparency, which intensifies competition and opens new possibilities for trade within the union. This would increase the spillover effects of fiscal policies. *Second*, EMU could lead to intense financial integration. This is already very visible in the bond market. The introduction of the euro has made it possible for a large euro-denominated bond market to emerge. What is the implication of financial integration for spillover effects of fiscal policies? The full integration of the bond markets implies that a fiscal policy action in one country, e.g., a higher budget deficit, increases the interest rate in the euro-bond market and therefore affects other countries. This effect is bound to be larger than when the bond markets are segmented. Thus, spillover effects through the interest rate channel increase.

The latter analysis has been influential in the design of the stability pact. The fear that excessive budget deficits and debt levels in one country would affect other countries through the interest rate channel has led to the idea that fiscal policies of the member states of the EMU should be tightly controlled.

One problem has been generally overlooked. Although spillovers through trade and interest rates are likely to increase in the EMU, the signs of the impact on economic activity through these channels are different. As a result, the *total* spillover of fiscal policies on the output levels of other countries may or may not increase as a result of the EMU. This can be easily shown in the context of a simple two-country Mundell–Fleming model. The intuition is the following. Take a fiscal expansion in, let us say, France. Because of stronger trade links, it increases output more than before the EMU in, let us say, Germany. At the same time, the French fiscal expansion has a higher impact on the long-term interest rate in Germany (through the unified euro-bond market) than prior to EMU. But this leads to a stronger negative effect on German output. It is unclear, therefore, whether the spillover effects of the French fiscal policy on foreign output will increase in EMU. One should also conclude that it is unclear whether a more intense coordination of fiscal policies in EMU will increase welfare.

Another way to phrase this result is as follows. The use of fiscal policies creates public good effects. Government spending in one country creates benefits in other countries. The benefits that spill over to other countries are likely to become more important in a monetary union. If, however, the expanding country has to bear full costs of financing this increase, it will engage in less government spending than it is desirable. Countries will have an incentive to spend too little.

However, at the same time, a monetary union increases the possibility to shift part of the financing costs of the expanding country to the other members of the union. This effect leads countries to increase their spending too much. Thus, both the benefits and the costs spilling over to the other countries tend to become more important. It is not clear, therefore, to what extent the monetary union distorts the incentives for individual countries to engage in too little or too much spending; nor is it obvious that one needs more coordination of fiscal policies in a monetary union.

b.   Asymmetric Shocks and the Need to Cooperate

The traditional theory of optimum currency areas has put a lot of emphasis on the notion of asymmetric shocks. This theory can be summarized as follows. Asymmetric shocks are likely to occur in a monetary union. It is therefore important to have an insurance mechanism that will allow individual nations to soften the blow of a negative shock on output and employment. Such an insurance mechanism can be provided by a unified European budget. By its very nature, such a unified budget automatically redistributes from countries experiencing good economic luck to countries experiencing bad economic luck.

If, however, no unified budget is set in place, which is the case in Euroland, the prescriptions of the theory are not favorable for the idea of cooperation. In the absence of a unified budget, countries have to use an intertemporal insurance mechanism to smooth asymmetric shocks, i.e., to use the budget deficit in a counter-cyclical way. This theory leads to the conclusion that in the absence of a unified budget, individual nations should use their fiscal policies to absorb asymmetric shocks. Therefore, national fiscal policies are tied to national objectives, making it very difficult to use fiscal policies in a cooperative way.

Recent research has stressed, however, that there is an alternative insurance mechanism for asymmetric shocks (see Asdrubali et al. 1990, Mélitz and Zumer 1998). It comes from financial integration. Empirical evidence suggests that integrated financial markets are capable of providing insurance against asymmetric shocks that is equally powerful as a unified budget. This insight has important implications for the EMU, which is likely to intensify financial integration. When financial markets are fully integrated, the EMU will function as an insurance mechanism against asymmetric shocks, reducing the need to provide insurance by budgetary transfers.

It is unclear, however, whether the insurance against asymmetric shocks provided by financial markets is satisfactory. The reason is that it insures the incomes of the relatively wealthy in a country hit by a negative shock, leaving the poor and the lower income groups with few financial assets unprotected. Modern nation states want to extend the insurance mechanism also to cover these groups. Consequently, a significant part of fiscal policies will continue to be tied up with the provision of insurance against asymmetric shocks in a monetary union.

8.   The Need for Cooperation between the ECB and the Budgetary Authorities

In the previous section we analyzed the need for cooperation between national budgetary authorities. In this section we study the need for cooperation between the ECB and the national governments setting fiscal policies. Here again the key concept is spillovers. Decisions made by the national budgetary authorities affect macroeconomic variables (inflation, output). Since the ECB is trying to control the same variables a spillover arises, and cooperation between the ECB and the fiscal authorities will generally improve decision making (and welfare).

There is no doubt that this is theoretically correct. The issue, however, is whether the size of the spillovers is sufficiently important to create significant welfare gains. Much of the literature on the desirable degree of cooperation between monetary and fiscal authorities has been based on models in which one monetary authority plays a game with one fiscal authority. In the context of EMU, however, one central bank faces twelve (and soon more) national fiscal authorities. This set-up weakens the case for cooperation significantly. The reason is that the spillover of budgetary policy of one country on the ECB is likely to be small. But differently, if one country, say France, follows expansionary budgetary policies, the impact of these policies on EMU-wide inflation and output is relatively small because France represents only about 20 percent of Euroland's output. In addition, France's actions are likely to be partially offset by other countries' actions. As a result, the French budgetary policies will interfere little with the ECB's policies. And the larger Euroland becomes, the weaker these spillover effects are, and the less the ECB has to worry about them. Thus, EMU is quite different from the United States, where the actions of the U.S. federal government are likely to have a significant effect on U.S. output and inflation, creating a need to coordinate monetary and fiscal policy.

There is another, political dimension that should be take into account. Decisions in the budgetary field are a prerogative of nationally elected parliaments. Decisions made by these bodies are slow and unpredictable. Attempts to coordinate these policies are, therefore, likely to be quite ineffective. In addition, systematic cooperation of fiscal policies is bound to reduce the power of national parliaments. It is unclear whether this is desirable in the absence of steps towards strengthening the democratic process at the EU level.

To conclude, the following points should be stressed. First, there is already a significant amount of coordination in the budgetary policy field in the context of the stability pact. The arguments developed here imply that there is little need to go beyond the stability pact and to enhance budgetary cooperation in a systematic way. This does not mean, of course, that occasional cooperation may be called for, for example, when strong enough common shocks occur.

Second, our conclusion that further steps towards budgetary cooperation are not needed does not imply that cooperation in other fields may not be desirable (e.g., tax harmonization, coordination of bank supervision).

9. Conclusion

The increased mobility of capital of the last few decades creates new challenges for the macroeconomic policies of the nation-states. In this paper, we have analyzed some of these challenges. One conclusion from our analysis is the following. Contrary to what is often alleged, increased capital mobility does not necessarily increase the need for coordination of monetary and fiscal policies. The reason is that this increased mobility of capital has led many nations to move towards greater exchange rate flexibility. And the latter reduces the need to cooperate in the monetary field. The effect on the need for fiscal policy coordination crucially depends on how spillovers of fiscal policies from one country to the other are changed. To the extent that capital market integration and trade integration go together, we do not know how the net spillovers of fiscal policies are affected.

Increased capital mobility creates many other challenges. We have analyzed several of these. We argued that while increased capital mobility puts more pressure on countries to move away form pegged exchange rates towards either more flexibility or more rigidity of the exchange rates, it also increases the temptation to escape this hard choice by reimposing capital controls. We argued, however, that one particular form of capital controls, i.e., the Tobin tax, is unlikely to succeed in giving countries a "Third Way" option.

Bibliography

Asdrubali, P., B. Sørensen, and O. Yosha (1996). Channels of Interstate Risk-Sharing: United States 1963–1990. *Quarterly Journal of Economics* 111: 1081–1110.

Barro, R., N.G. Mankiw, and X. Sala-i-Martin (1992). Capital Mobility in Neoclassical Models of Growth. In *NBER Conference on Economic Fluctuations*. Cambridge, Mass.

Bergsten, C.F., and J. Williamson (1983). Exchange Rates and Trade Policy. In W. Cline (ed.), *Trade Policy in the 1980s*. Washington, D.C.: Institute of International Economics.

Bordo, M.D., B. Eichengreen, and D.A. Irwin (1999). Is Globalization Today Really Different than Globalization a Hundred Years Ago? NBER Working Paper 7195, Cambridge, Mass.

De Grauwe, P. (2000). Controls on Capital Flows and the Tobin Tax. Center of Economic Studies Discussion Paper 00.02. Leuven.

De Grauwe, P., H. Dewachter, and Y. Aksoy (1999). The ECB: Decision Rules and Macroeconomic Performance. CEPR Discussion Paper 2067. London

Eichengreen, B., and C. Wyplosz (1993). The Unstable EMS. *Brookings Papers on Economic Activity* (1): 51–124.

Frankel, J. (1991). Quantifying International Capital Mobility in the 1980s. In B.D. Bernheim and J.B. Shoven (eds.), *National Saving and Economic Performance*. Chicago: University of Chicago Press.

Fratianni, M., and J. von Hagen (1990). The European Monetary System Ten Years After. In A.H. Meltzer and C. Plosser (eds.), *Unit Roots, Investment Measures and Other Essays*. Carnegie-Rochester Conference Series on Public Policy, No. 32.

Ghosh, A., A.M. Gulde, and H.C. Wolf (1998). Currency Boards: The Ultimate Fix? IMF Working Paper 98/8, Washington, D.C.

Jansen, W.J. (1996). The Feldstein–Horioka Test of International Capital Mobility: Is It Feasible? IMF Working Paper WP/96/100, Washington, D.C.

Jones, M.T., and M. Obstfeld (1997). Saving, Investment, and Gold: A Reassessment of Historical Current Account Data. NBER Working Paper 6103, Cambridge, Mass.

Lawrence, R., A. Bressand, and I. Takatoshi (1996). *A Vision for the World Economy: Openness, Diversity and Cohesion*. Washington D.C.: The Brookings Institution.

Lyons, R. (1999). The Microstructure Approach to Exchange Rates. Preliminary version available at http://haas.berkeley.edu/~lyons/NewBook.html.

McKibbin, W.J. (1997). Empirical Evidence on International Economic Policy Coordination. In M.U. Fratianni, D. Salvatore, and J. von Hagen (eds.), *Macroeconomic Policy in Open Economies. Handbook of Comparative Economic Policies*. Volume 5. Westport, Conn.: Greenwood Press.

Mélitz, J., and F. Zumer (1999). Interregional and International Risk Sharing and Lessons for EMU. CEPR Discussion Paper 2154. London.

Mussa, M., P. Masson, A. Swoboda, E. Jadresic, P. Mauro, and A. Berg (2000). Exchange Rate Regimes in an Increasingly Integrated World Economy. IMF Occasional Paper 193, Washington, D.C.

Obstfeld, M. (1986). *Capital Mobility in the World Economy: Theory and Measurement.* Carnegie-Rochester Conference Series on Public Policy, No. 24.

Radelet, S., and J.D. Sachs (1998). The East Asian Financial Crisis: Diagnosis, Remedies, Prospects. *Brookings Papers on Economic Activity* (1): 1–90.

Rodrik, D. (1998). The Debate Over Globalisation: How to Move Forward by Looking Backward. Mimeo, Harvard University, Cambridge, Mass.

Sachs, J., and A. Warner (1995). Economic Reform and the Process of Global Integration. *Brookings Papers on Economic Activity* (1): 1–118.

Summers, L.H. (1998). Tax Policy and International Competitiveness. In J. Frenkel (ed.), *International Aspects of Financial Policies.* Chicago: University of Chicago Press.

Taylor, A.M. (1994). Domestic Savings and International Capital Flows Reconsidered. NBER Working Paper 4892. Cambridge, Mass.

— (1996). International Capital Mobility in History: The Saving-Investment Relationship. NBER Working Paper 5743. Cambridge, Mass.

Vamvakidis, A., and R. Wacziarg (1998). Developing Countries and the Feldstein–Horioka Puzzle. IMF Working Paper 98/2. Washington, D.C.

Williamson, J. (1985). *The Exchange Rate System: Policy Analyses in International Economics.* Washington, D.C.: Institute of International Economics.

Zevin, R. (1992). Are World Financial Markets More Open? If So, Why and with What Effects? In T. Banuri and J.B. Schor (eds.), *Financial Openness and National Autonomy.* Oxford: Clarendon Press.

p 177;

Comment on Paul De Grauwe and Magdalena Polan

Jean Pisani-Ferry

The paper I am commenting on covers a wide range of issues in a selective fashion. I thus feel free to follow the same selective pattern, which in my case means skipping topics I agree on with the authors and focusing on those on which my views may differ from theirs. There is much I agree on with the authors of this stimulating and subtle paper. My list of topics for discussion is the following:

1. Capital mobility.
2. Exchange rate in emerging economies.
3. The Tobin tax.
4. Fiscal policy coordination in EMU.

1. Capital Mobility

Paul De Grauwe and Magdalena Polan conclude their brief survey of the issue by stating that "the degree of financial market integration today is not dramatically higher than one hundred years ago," and go on to write that "there seems to be evidence that [it] has been increasing in the last few decades" (p. 178). I agree with both statements (although the second is in my view a clear understatement), but it seems to me that they miss an important point.

I will not dispute that capital mobility was high under the gold standard. What is striking is that the composition of flows was significantly different: although gross flows, as recorded by FX markets statistics, are now several times higher than at the beginning of the century, net flows (i.e., net saving flows from surplus to deficit regions) are probably lower, as suggested by the correlation between saving and investment. Evidence for this view was provided in a paper by Tamim Bayoumi (1999) for a previous Kiel Week Conference, which shows that although the β coefficient on the saving ratio in Feldstein–Horioka equations has dropped from 0.85 to 0.60 in recent years, it was insignificant under the gold standard (as it is within countries, see, for example, Bayoumi's estimates on Canadian provinces).

In spite of a dramatic move towards capital markets liberalization, as illustrated by the number of countries which have gone all the way to financial convertibility, the world economy has thus not yet been able to reach a stage in which investment is totally insensitive to the geographical origin of savings.

The world economy has thus not reaped the full benefits of financial globalization. At the same time, countries which had lifted capital controls have frequently experienced financial crises and endured severe costs. It is this contrast between the costs and the benefits side which fuels claims that capital account liberalization was a bad idea in the first place, and that any comparison between the benefits of trade and financial liberalization is deeply flawed (Bhagwati 1998). And although recent crises may have led governments to focus more on meeting the preconditions to a successful financial liberalization (for example, the robustness of the financial system or the adequacy of the exchange rate regime), they have also encouraged policymakers to lower the ceiling on what they consider as risk-free current account deficits, thereby reducing the potential benefits of liberalization alongside the costs.

This could be more a matter for concern than suggested by the authors, as support for financial liberalization ultimately depends on the (perceived) balance between costs and benefits.

2.   Exchange Rate Policies in Emerging Economies

The paper by and large embraces the new consensus view on exchange rate regimes, which essentially states that there is evidence of a polarization towards either of the two "corner solutions" (free float, and currency board or dollarization/euroization), and that this is an optimal response to the crises of recent years. Both statements deserve qualifications:

a. The evidence on polarization is much weaker than suggested by official policy formulations, as recorded in the IMF's data. The picture that emerges from an analysis of monthly volatility data vis-à-vis the major currencies provides evidence of de facto anchoring policies, even for countries officially in a pure floating exchange rate regime (Bénassy-Quéré and Coeuré 2000).

b. The lesson from recent crises is more subtle than the frequently held view that intermediate regimes between a pure float and a currency board are a recipe for disaster. From the (indisputable) assertion that rigid pegs without a strong institutional backing are a recipe for crisis, it does not follow logically that a choice should be made between a free float and relinquishing monetary sovereignty altogether. Intermediate regimes—i.e., regimes in which the weight attached to the exchange rate objective in the central bank's loss function is neither zero nor one—can be viable, provided they meet two conditions. First, they should leave room for deviation from the exchange rate target, in order to let private agents learn that the central bank is willing to accept those deviations, and price the risk accordingly. Second, they should be transparent, i.e., the central bank should let it be known how it would react to an appreciation/depreciation.

This is not do deny that referring to the corner solutions can be useful for clarification purposes, nor that over the long run, some polarization has occurred as a consequence of financial liberalization. However, academics should be careful and avoid giving the impression that there is a consensus to ban intermediate solutions. They should rather highlight the conditions for such regimes (which continue to have appeal for governments which are not entirely indifferent to exchange rate developments and are therefore not really willing to adopt a clean float strategy) to be viable. Exchange rate policies have already suffered enough from excessively volatile fads.

3.   The Tobin Tax

It may be a surprise for non-European participants in this conference that De Grauwe and Polan devote several pages to discussing why a Tobin tax would not reduce volatility in exchange markets. There is nothing I disagree with in their demonstration, which rightly draws on the new literature on the microstructure of forex markets. However, I doubt that it can convince the (numerous) proponents of the Tobin tax, because the appeal of this proposal is more a political than an economic one. Economists should thus devote more attention to understanding why, in the first place, proposals of this sort keep their political attraction after it has been convincingly demonstrated that they could not reach the goal they are supposed to serve.

The appeal of the Tobin tax is in my view that it is pictured as an efficiency-enhancing device that would stem destabilizing speculation while promoting tax fairness and redistribution to the poor. It thus looks, in a way, like an economist's dream measure which would promote both efficiency and equity. To the general public, it has, in addition, the appeal of being modern, because it would not ban nor regulate exchange market speculation, but only subject it to taxation, like any other economic activity. The Tobin tax is thus a very clever product of political marketing, and the success it has received in several European countries should come at no surprise.

The lessons for economists is that they should not limit themselves to demonstrating why a Tobin tax would not reach the goal it is deemed to reach. This "it won't work" critique is a political nonstarter, because it can only be met with skepticism by the public. Economists should, rather, take up the broader issue of efficiency-and-equity-enhancing reforms and make alternative proposals; that would force the Tobin tax proponents to enter into a serious discussion on the merits of their own favorite measure.

4.   Fiscal Policy Coordination in EMU

This last topic is the only one I seriously disagree on with the authors. De Grauwe and Polan argue that as monetary union increases both the positive spill-over effects of fiscal policy (through goods markets integration) and the negative ones (through capital markets integration), the resulting net effect is indeterminate. For that reason, there is in their view no strong case for coordinating fiscal policies in a monetary union.

The argument is obviously correct in a technical sense, although to demonstrate that an issue can only be settled at the empirical level (as opposed to the theoretical one) does not demonstrate that it is irrelevant. But it misses an important point. There are strong arguments for fiscal policy coordination within the euro zone which do not depend on a particular assumption on the sign of spill-over effects. I would like to mention three which in my view carry some weight:

a. The case for coordination is first of all a political one. As Paul De Grauwe eloquently argued in a *Financial Times* article (De Grauwe 1998), the ECB can only rely on a strong legitimacy if ministers are able to talk to each other and to take collective responsibility for what belongs to governments (and not to the central bank) : fiscal policies, structural policies and tax policies, to mention only the three major aspects. An ECB which would be perceived as acting in isolation would be a weak central bank, and would lack legitimacy.

b. As monetary and fiscal policy are to a certain extent substitutable, there is something like a policy mix for Euroland as a whole. Determining the corresponding stance may call for coordination between governments.

In a world in which monetary policy would only affect prices and fiscal policy only real variables, the notion of a policy mix would be irrelevant, since the effects of both instruments could be separated. However, this is only true in the very long run, and for most practical purposes, monetary and fiscal policy are to a certain degree substitutable. This means that there are different possible responses to a shock, and that the optimal fiscal policy response depends on the assumptions made about monetary policy (and vice versa). It is thus necessary that some mechanism exist for coordinating fiscal policies, and that provisions be made for a dialogue between fiscal and monetary authorities. This can be achieved either through the setting of policy rules or through a more discretionary approach, but what is important is that the mere existence of a common monetary policy creates a motive for coordination that did not exist prior to monetary union.

The relevance of this issue was highlighted in the policy debates of 1998–1999 on the occasion of the Asian crisis, as policymakers in Europe discussed whether the impact of this negative demand shock should be mitigated through fiscal or monetary policies. Whatever the actual outcome of such discussion, the corresponding mix could only be reached through an implicit or explicit coordination between fiscal policy authorities.

c. The common inflation rate, the common exchange rate, and the common current account balance are collective goods which are specific to a monetary union. As they are treated (obviously to a different extent) as policy objectives by the central bank and can at the same time be affected by fiscal policies, this creates a channel of interdependence which is both specific to EMU and unambiguous in sign.

The inflation case is probably the most straightforward. As the ECB has clearly stated that it takes as a target variable the average Euroland inflation rate, an increase in a participating country's inflation directly impacts on its neighbors through the reaction its effects on Euroland inflation trigger from the ECB. As fiscal policy in an EMU country affects its own prices in the short run, a fiscal policy expansion in a member country increases the common inflation rate and thus triggers a tightening by the ECB. This effect is unambiguous in sign and does not depend on any particular assumption on goods and capital market spillovers, as it *only* arises from the definition of its objective by the ECB. In fact, it would remain even in the absence of *any* trade and capital flows between participating countries.

Divergence in inflation rates within EMU (to the extent they do not correspond to equilibrating mechanisms, such as Balassa–Samuelson effects) thus give rise to policy discussions on the appropriate response to price developments in countries with higher inflation. Those countries are adamant that monetary policy should be tightened. Low inflation countries naturally insist that localized inflationary pressures should be countered through country-specific fiscal policy actions rather than monetary policy actions that would affect all member states in an undiscriminatory way. Reconciling this divergence of views is precisely the purpose of coordination.

The case for coordination is thus stronger than suggested by De Grauwe and Polan. Obviously, there are also well-known obstacles to coordination, which deserve discussion in the context of EMU. Such a discussion should, however, take as a basis that the challenges to be addressed are specific and that the creation of a monetary union thus calls for a specific approach to coordination.

Bibliography

Bayoumi, T. (1999). Is There a World Capital Market? In H. Siebert (ed.), *Globalization and Labor*. Tübingen: Mohr Siebeck.

Bénassy-Quéré, A., and B. Coeuré (2000). *Big and Small Currencies: The Regional Connection*. Centre d'études prospectives et d'informations internationales (CEPII) Working Paper 2000–2009, Paris.

Bhagwati, J. (1998). The Capital Myth. *Foreign Affairs*, May–June, pp. 7–12.

De Grauwe, P. (1998). Law unto Itself. *Financial Times*, November 12.

Tommaso Padoa-Schioppa

G21 G28

F32

Increased Capital Mobility:
A Challenge for the Regulation of Capital Markets

1.   Introduction

The way in which policy-makers look at capital flows, on the one hand, and financial markets, on the other, has changed significantly over the past 50 years. The Bretton Woods system and the current global financial system are so remote from one another in this regard that, for the sake of simplicity, they will be referred to hereinafter as the "Old World" and the "New World."

Just as the Old World suffered from contradictions and was faced with a number of challenges in terms of economic policy, so too does the New World at the current juncture. Completely different, however, is the nature of such contradictions and challenges. As argued in Section 2 of this article, the New World is no longer exposed to the macroeconomic inconsistencies that eventually caused the collapse of Bretton Woods. The New World also provides for a better allocation of resources than in the past and represents a departure from those paradigms of "financial repression" which, in the Old World, sheltered domestic markets from competitive pressure.

The free movement of capital, while being a key factor underlying these positive achievements, implies a new challenge for economic policy. The challenge concerns the *stability* of the global financial market, in both its systemic and local components. As a result, the policy emphasis in the New World has shifted from controls on capital flows to prudent risk management, from ex post crisis resolution to what is often referred to as crisis prevention, and from uncoordinated measures at the country level to the beginning of an integrated set of internationally agreed measures.

In dealing with these issues, the official community is gradually constructing a new *institutional* framework. As will be explained in Section 3, this is neither the

---

Remark: The author is grateful to Ettore Dorrucci and Panagiotis Strouzas for their helpful assistance in the preparation of this article.

outcome of a formal ad hoc conference nor the manifestation of a top-down government of economy, as in the case of Bretton Woods. Rather, it is an informal process triggered by continuous dialogue between policy-makers and their market constituency. The unstructured nature of the present institutional arrangements may induce observers to define the New World as a *nonsystem*. However, such arrangements are, on the contrary, a key component of a system that can be defined as a "market-led international monetary system" (or M-IMS), as opposed to the preceding "government-led international monetary system" (or G-IMS).[1]

Much progress has been made in improving the M-IMS. The list of new initiatives is quite impressive, even if we consider the last two years alone. These initiatives range from G7 action to the creation of the Financial Stability Forum, from areas such as "accounting" and "offshore centers" to the new Basel rules. They concern either the institutional framework of the M-IMS or the new conceptual approach required for policy-making in the area of financial stability. While acknowledging the importance of the latter aspect—which, to a large extent, is still open to discussion—the main aim of the following few pages is to highlight those unresolved issues pertaining to the institutional framework of the M-IMS which need to be tackled in the medium term.

2. Why a Challenge?

In the "extreme" Bretton Woods world of the early 1950s, capital flows passed mainly through official channels, and each country could regulate and supervise the national financial system in its own way. The free movement of capital was gradually reestablished between the late 1950s and the 1960s, as an implication of the liberalization of trade, which required the restoration of currency convertibility to allow the settlement of current account transactions. Convertibility, in turn, allowed market participants to exploit trade-related transactions (e.g., so-called leads and lags) as a way of moving capital across borders. Technology, and financial innovation made it increasingly possible to circumvent restrictions. The formal relaxation of capital controls was induced by their declining effectiveness, but also by a growing recognition of the positive role markets play in the international allocation of savings.

Following the end of Bretton Woods, a long transition began during which elements of the Old World coexisted with elements of the emerging New World.

---

[1]   The distinction between M-IMS and G-IMS was introduced by Padoa-Schioppa and Saccomanni (1994: 236).

As early as the 1970s, major industrial countries, such as the United States, Germany and, at the end of the decade, the United Kingdom, handed over to the markets—either de jure or de facto—not only the determination of the exchange rate, but also the international allocation of savings. However, until the early 1990s, permanent and/or temporary restrictions on capital movement continued to be regarded, albeit with decreasing conviction, as a key policy instrument, not only for developing countries, but also in industrial countries. France and Italy only completed the process of liberalization in 1990. Spain, Portugal, and Ireland made use of temporary restrictions during the 1992 crisis. In the European Union, the free movement of capital *erga omnes* became a basic rule directly enforced by the underlying treaty only as of January 1, 1994.

In the course of the 1990s, the balance of power clearly shifted from governments to the markets. The free movement of capital has thus become the only conceivable regime in the industrial world, and a fundamental standard to be applied—if not already in place—by emerging and transition economies as well.

As long as restrictions on capital flows played a role in the industrial world, they were designed mainly for the achievement of two basic objectives: the *autonomy* of monetary and exchange rate policies, and the *stability* of the domestic financial system.

Restrictions gave a degree of *autonomy* to national authorities in deciding—in the context of free international trade—the "price of their domestic currency," expressed in terms of both another currency (exchange rate) and its use at different points in time (domestic interest rates). As such, restrictions were the instrument through which countries tried to solve the "inconsistent quartet," i.e., the conflict between free trade, the free movement of capital, fixed exchange rates and monetary policy autonomy.[2] It is no surprise that the system became intrinsically inconsistent when the effectiveness of capital controls decreased. The final breakdown of the fixed exchange rate regime in September 1973 was also the result of enhanced capital mobility, since the amount of time allowed for countries with deficient fundamentals to contemplate policy options before market pressure became unsustainable had become increasingly shorter.

In the New World, capital restrictions are no longer regarded as the appropriate way to solve the "inconsistent quartet." At the *global* level, the adoption of

---

[2] The expression "inconsistent quartet" was first used in 1982 by Padoa-Schioppa in his speech entitled "Mobilità dei capitali: perché la Comunità è inadempiente" held at the "Secondo simposio di banche europee" conference (Milan, June 1982), published in Padoa-Schioppa (1992). As Henry C. Wallich noted as early as 1972, in the Bretton Woods system the inconsistency described above was well known to economists, but not sufficiently acknowledged in the institutional framework or in the basic principles for policy-making (Wallich 1972).

exchange rate arrangements based on free or managed floating has prevailed. At the *regional* level, experience has shown that the more integration moves from mere free trade towards a single market, or even an economic union, the greater the need for intra-area exchange rate stability and, eventually, irrevocably fixed exchange rates. This comes into conflict with the autonomy of monetary policy at the national level. In Europe, the need for monetary union was eventually recognized as the preferred solution.

From a financial *stability* point of view, restrictions on both movements of capital and cross-border financial services were aimed at artificially isolating the domestic financial system from international sources of instability and outside competition. As such, restrictions had not only a prudential, but also a protectionist function. "Financial repression" was usually justified by the inadequate development of domestic financial markets (in terms of size, liquidity, competition among intermediaries, etc.). This approach, however, revealed at least two weaknesses. *First*, in pursuing the objective of stability (also) via capital controls, these restrictions undermined another objective—efficiency—to the extent that markets were only allowed to determine the international allocation of resources for specific items of the balance of payments (e.g., goods and nonfinancial services). *Second*, experience has shown that financial repression, if it is not linked to a path of liberalization, eventually produces more problems than solutions, even from a financial stability point of view, since it is one of the main sources of domestic fragility. This explains why, in the Old World, the authorities of industrial countries were continuously confronted with the need to find a balance between interference with and the liberalization of capital flows and financial markets.

Could "financial repression" be applied, in the New World, to those emerging and transition economies which have not yet completed the liberalization of the movement of capital? The answer to this question is "no"; there is, however, a caveat.

The caveat is that, as experience shows, the increase in market confidence in a country often gives rise to massive short-term capital inflows, which eventually results in real exchange rate appreciation. Currency appreciation, in turn, produces a misallocation of resources, which damages the overall economy of the country until sudden reversals in the flows precipitate a currency crisis. In order to help to avoid this chain of events, temporary market-based controls on short-term capital inflows could be envisaged if coupled with a carefully designed sequence of liberalization measures. This may also improve the maturity composition of foreign debt. The Chilean authorities adopted these kinds of measures between 1991 and 1998, when they raised the cost of short-term external financing, mainly by imposing an unremunerated reserve requirement of one year on portfolio inflows. Supplemented by the liberalization of current account transactions and capital outflows, this strategy has proved successful in preserving the peso from real appreciation in a context of tight domestic monetary policy.

More fundamentally, however, it is undeniable that permanent capital controls distort the efficient allocation of resources. This is particularly true in a world where the financing of emerging markets depends heavily on the inflow of private capital from abroad. According to estimates by the Institute of International Finance, in the 1990s 85 percent of total net inflows of capital to the major borrowing countries came from the private sector. Such inflows averaged $170 billion per annum, compared with $35 billion per annum in the 1980s. Although they still play a crucial role for crisis-hit countries, net official flows averaged just $30 billion per annum throughout the 1990s and reached only $5 billion in 1999, their lowest level for over 20 years.

While, in the New World, the free movement of capital is allowed to operate as a mechanism for pursuing efficiency, its implications for *stability* are more challenging. The risk of financial instability may, indeed, be heightened by the interaction of two factors.

The first factor is *inadequate supervision* in most emerging and transition economies. The increased opportunities for portfolio diversification by global investors, as well as the heavy dependence on external finance by emerging and transition economies, significantly increase the flow of funds towards countries where the institutions and practices designed to ensure financial stability are not yet well developed. When foreign investors distrust the safety and soundness of their counterparts, a sudden reversal of capital flows may easily affect the ability of local financial institutions to honor their obligations. In the absence of internationally agreed and strictly enforced standards, the inevitable imperfections of tools aimed at limiting the build-up of risk exposures can produce adverse effects on the confidence of investors, thus inducing "flight-to-quality" behavior.

The second factor is *contagion*, i.e., the international transmission of instability. In this respect, the New World has to deal with a number of market failures, which may be exacerbated by a lack of transparency. Financial markets are prone to demonstrate a herd mentality, particularly when the amount of public information available for distinguishing between "good" and "bad" borrowers is scarce. Experience shows that there has been a lack of stabilizing speculators for emerging and transition economies, resulting in very volatile capital flows. It is, therefore, not true that "there are no innocent victims": in an epidemic the weakest are, of course, affected more severely than the strong, but if there were no epidemic, the weak would not die. Needless to say, this argument should never become an easy way for policy-makers to shirk responsibility.

These two factors are the basic reasons why the free movement of capital constitutes a new challenge for economic policy.

The focus of economic policy in dealing with cross-border financial activity has consequently changed in the post-Bretton Woods era, particularly over the past decade. In particular, the shift from the Old to the New World has entailed three main implications for policy-makers.

First, the regulation of the movement of capital has ceased to be a key policy tool, and has been replaced by a set of instruments which are designed to promote *prudent risk management* of assets and liabilities by market players. Indeed, the crises which occurred in the second half of the 1990s have highlighted the importance of sound risk management on the part of market participants, including governments and central banks, in terms of helping countries to withstand shocks and providing support when adjustment is necessary. Sound market behavior is encouraged through a wide range of policy tools, extending from market regulation to surveillance and supervision.

Second, the emphasis is increasingly shifting from crisis management to what is known as *crisis prevention*[3] and *private sector involvement.* When capital flows were mainly channelled through the banking system (which was the case even in the early stages of the New World), public authorities were in a better position to promote negotiated solutions among the parties involved in a financial crisis. Today, capital flows are increasingly centered on capital markets and institutional investors. The more fragmented the nature of investors and the more diversified the financial instruments used, the greater the capacity of lenders to "vote with their feet" and the less viable an ex post workout of crises.

Third, the scope for *international cooperation* has become much larger than in the past, since the relevant environment for all market participants, even the most domestically rooted, is the global financial system. It is no coincidence that the first initiative to create an international "standard-setter" in the financial area—the Basel Committee on Banking Supervision (or Basel Committee)—was taken in 1974, immediately after the end of Bretton Woods.[4] The need for cooperation has increased dramatically since then. The ultimate solution in this respect would be a globally harmonized regulatory framework. Today, however, such a radical

---

[3]   Of course, this does not mean that crisis management is no longer an issue. While this paper does not deal with the subject, crisis management remains a key aspect even in the M-IMS, not least because crisis prevention depends, to a large extent, on the way a potential crisis is managed. Moreover, to call anything which is not crisis management "crisis prevention" would be too narrow a view, since regulation is not only "crisis-related."

[4]   Specifically, the foreign-exchange-related collapse of Bankhaus Herstatt in 1974 caused the G10 central banks to establish a "Committee of Banking Supervisors" in Basel, under the aegis of the Bank for International Settlements (BIS), to examine the international implications of the crisis.

solution would be unworkable, since it would come into conflict with the diversity of currencies, legislations, fiscal regimes, cultures, and, last but not least, the degree of development of each financial market.

Almost inevitably, a world divided into sovereign states has moved more rapidly towards full capital mobility than towards common regulation. The asymmetry between the free movement of capital, which takes place at a global level, and policy-making, which remains mainly in the national domain, can have implications for financial stability and calls for enhanced international cooperation.

Enhanced cooperation, the corrective for the absence of a single authority, presents a number of challenges with regard to the *institutional* framework within which the international community aims to achieve stability.

3.   The Institutional Framework

If one had to identify a single event after which global financial stability became a *fundamental policy objective* to be pursued on an *ongoing* basis, it would be the crisis in Mexico in 1994–1995 and the ensuing G7 Halifax meeting in June 1995. Many important results had already been achieved before Halifax (e.g., the Basel Capital Accord and the so-called Lamfalussy standards[5]), but mainly as a result of ad hoc "crisis-driven" initiatives at the technical level. Since Halifax, topics related to financial stability have regularly been included on the agenda of top-level fora such as the G7, thus benefiting from political guidance. Moreover, financial issues have been incorporated more systematically into the standard-setting and surveillance work carried out by the Bretton Woods institutions, the IMF and the World Bank.

Although Mexico/Halifax was not the start of the process, a new permanent institutional framework has increasingly taken shape since then. While it is far from complete, this framework has already become a pillar of the M-IMS.

The new global framework is based on a "soft mode," as opposed to the "hard mode" pursued, at the regional level, by the European Union (EU). In the *soft mode*, decisions are not legally binding. Effectiveness is rooted in both consensus and a "desire to be respectable." The soft mode is not supranational—as is the case with hard-mode EU institutions—but international in nature. Decisions are taken and implemented on a voluntary basis by the participating national authorities, which are the only entities involved at all levels of the decision-making

---

5   The Lamfalussy standards apply to cross-border and multicurrency netting schemes.

process. This is the prevailing approach of the New World, since it reconciles diversity with the need for harmonization. Two simple, interrelated instruments contribute to the functioning of this method. First, *best practices*: instead of setting a standard applicable in the same way everywhere, the international community agrees on a kind of benchmark solution, giving the national authorities the freedom to adhere to such a benchmark in a manner which is tailored to local characteristics. Second, *best endeavors*: since national authorities are not assigned the same tasks and responsibilities, they pursue common objectives "on a best effort basis."

Conversely, with the *hard mode*, European institutions have not only extensive powers, but even exclusive competencies in many fields. Community legislation is binding for EU member states. In the financial sector, harmonization of the regulatory framework has progressed as far as it could, albeit respecting the peculiarities of national systems. The EU can, therefore, be considered an exception in the M-IMS, although its original formula may become a point of reference for other regions of the world.

The current configuration of the soft mode is based on three features. The first feature concerns the allocation of responsibilities in the *regulation and supervision* of financial markets. Responsibilities are currently shared in accordance with the following informal arrangement.

The *political impulse* is provided by fora such as the G7 and, more recently, the G20 and the Financial Stability Forum (FSF). In Halifax, for instance, the G7 gave the political input without which the "Core Principles for Effective Banking Supervision" of the Basel Committee would not have been possible. This development triggered the adoption of similar approaches by other international fora.

Rule-making is based on a twofold approach. At the international level, *standard-setting* for financial intermediaries[6] is left mainly to the existing specialized fora: the Basel Committee, the International Organisation of Securities Commissions (IOSCO), and the International Association of Insurance Supervisors (IAIS). The work of these fora is complemented by that of the G10 Committee for Payment and Securities Settlement Systems (CPSS), the International Accounting Standards' Committee (IASC), and, last but not least, the International Monetary Fund (IMF). The main features and achievements of all these bodies, apart from the IMF, are presented in Appendices 1 and 2, respectively. At the local level, *financial regulation*, which basically applies the internationally agreed standards, is assigned to national authorities.

---

[6]   Standard-setting can, of course, also apply to the public and corporate sectors.

Rule enforcement is the joint outcome of the *surveillance* carried out by the IMF and the World Bank at the international level and *supervision* at the national level.

The second feature is that no "statutory authority" for the *liberalization* of *capital movement* exists in the present configuration. While the liberalization of cross-border economic activity is the main objective of the World Trade Organization (WTO), its mandate only covers current account transactions. Similarly, the IMF's Articles of Agreement relate to current transactions, but not to the financial account. Comprehensive regimes for capital flows have only been agreed for arrangements involving limited groups of countries such as the EU and the members of the Organization for Economic Cooperation and Development (OECD). No global regime has emerged as yet. This implies that regulation and liberalization are not—as they should in principle be—part of a single process, as is in fact the case for the European hard mode.

Several approaches have been put forward as potential solutions to this contradiction. A first option would consist of amending the IMF's Articles of Agreement in order to entrust it with jurisdiction over financial account transactions. This solution, however, may be criticized on the basis of the principle according to which liberalizing capital movements and handling with the problems raised by it should remain separate tasks entrusted to separate authorities. While the latter task certainly falls within the domain of the IMF, the former might, according to an alternative option, be entrusted to the WTO, the competencies of which would thus be broadened so as to comprise all balance-of-payments transactions. This would make sense in a world where the instruments of financial repression are no longer aimed at macroeconomic equilibrium, but rather have a micro-protectionist function.

The third feature of the current configuration is that *the country composition of international fora and organizations* is different among and within the aforementioned three levels of international policy-making (i.e., the political impulse, rule-making, and rule enforcement). The political impulse mainly comes from the countries participating in the G7, although the G20 could gain importance over time. In setting standards for the banking industry, the Basel Committee reports to the G10, whereas membership of IOSCO and the IAIS is extended to all countries. The enforcement of rules at the international level is entrusted to the IMF and the World Bank, which are also worldwide institutions. These discrepancies are a source of complication, as they could result in friction between and within the various levels of international policy-making.

4.   Four Challenges

The practicality of the soft mode is an essential condition for achieving the effective functioning of the global financial system, but practicality alone is not sufficient. We are still faced with a number of major gaps. Although the list is by no means exhaustive, the following four challenges should be highlighted, which the international community may be able to address in the medium term.

The first challenge is to *harden the soft mode* by enhancing, if required, the functioning of international standard-setters.

The *Basel Committee* has gradually developed a very effective working method, and is probably the benchmark for the functioning of the soft mode. Its achievements rely on a number of specific features which are not entirely shared by other fora. For instance, as mentioned above, the initiatives of the Basel Committee are promoted by a small group of leading countries (G10 area), with their central banks providing substantial support. At the same time, these countries operate at the top of an extensive network of regional committees, which, in turn, are capable of linking each country with the others. An effective "transmission mechanism" is therefore in place. In this context, the Basel Committee also provides training programs, which facilitate the exchange of ideas and principles. Finally, the Basel Committee can count on significant resources and has a strong secretariat which is heavily involved in the drafting of relevant papers.

IOSCO's mandate is to ensure an improved regulation of securities markets by establishing international standards and monitoring their implementation. Its general meetings take place once a year in the form of an assembly only. IOSCO does not benefit from a direct reporting line to a strong forum responsible for the political impulse, such as the G10 in the case of the Basel Committee. The Technical Committee of IOSCO aims at more fruitful cooperation along the organizational lines of the Basel Committee, while the establishment of regional committees has contributed to enhancing work in accordance with the principles of the soft mode.

The *IAIS*, established in 1994, is a young organization supported by a small secretariat. It aims to improve the supervision of the insurance industry and to develop sound regulatory standards for its members. As with IOSCO, its main convention is its Annual Conference. An additional element which could have an impact on the application of commonly agreed standards in this area is the fact that, in the United States, insurance supervision is not coordinated at the federal level, but is, rather, conducted by individual authorities in each of the 51 states.

The *CPSS* was set up in 1990; its predecessors,[7] however, have been at work since 1980. It mainly serves as a forum for the G10 central banks to set standards in order to reduce risks in payment and settlement systems. To this end, it monitors and analyzes developments in domestic payment, settlement, and clearing systems, as well as in cross-border and multicurrency settlement schemes. It also coordinates the oversight functions assumed by the G10 central banks in respect of payment systems. The Committee has increasingly extended its work outside the G10 area. Non-G10 central banks are more and more frequently associated with initiatives of the Committee and of its subgroups. The CPSS also cooperates with other standard-setters, and maintains contacts with the most important payment system providers and industry associations.

Accounting is another important area in which international rules are needed. This would allow for a comparison of banks' capital adequacy across jurisdictions. In this area, "competition between rules" is the mode still prevailing, which limits the scope for action by the relevant rule-maker, the *IASC*. Indeed, the main tasks of this committee have so far consisted of, first, selecting the best accounting standards available at the national level, and second, seeking to achieve worldwide acceptance of these standards. Some very recent developments could, however, enhance the role of the IASC in the near future.[8]

Finally, the *IMF* sets standards which are helpful for enhancing its surveillance functions, i.e., mainly statistical standards—including those relating to the financial sector—and "codes of good practices" for policy-making in areas, such as monetary and fiscal policy, which have a bearing on financial stability. This raises the question, discussed in the following section, of whether a conflict could arise between the different roles played by the IMF as, on the one hand, a rule-maker and, on the other, a rule-enforcer.

Beside ad hoc adjustments which may be required to enhance the functioning of each international standard-setter, another key issue is to improve the method and logistics of cooperation in compliance with a set of good practices shared by all bodies. The following are a few concrete suggestions. First, with regard to the number of participants in each body—i.e., the number of people present around the table—20–25 participants is the maximum number which allows for an interactive "conversational mode" and a cross-fertilization of minds. Second, the adoption of a single working language is crucial. Third, the duration of the chair-

---

[7] The "Group of Experts in Payment Systems" and the "Ad-hoc Committee."

[8] In May 2000 a committee chaired by Paul Volcker nominated trustees with the mandate to transform the IASC into a body capable of promoting new standards and coordinating national practices.

person's term is also very important, as work continuity cannot be achieved if the chairperson changes too frequently. Fourth, the soft mode calls for an active role to be played by the secretariat of each relevant body: in particular, the issue of "who" (either the secretariat itself or the chairperson) is responsible for drafting documents is of great practical relevance.

The second challenge consists of *enforcing the rules*. At the international level, the key institution in this area should remain the IMF. The IMF has statutory power in the framework of both Article IV consultations and conditional support to national programs. Its coverage is worldwide, and it has the necessary resources and expertise to carry out its assessments at both the micro and the macro level. Finally, it comprises a strong policy body in the form of the Executive Board.

Rule enforcement by the IMF would be facilitated to the extent that prior rule-making could be carried out by fora in which competent national authorities are represented. There is a simple reason for this: in the soft mode—i.e., in a world where each achievement is the outcome of voluntary collaboration—the more co-operative and even laborious the process to achieve an agreement on new rules, the more successful the agreement, and vice versa. If a given standard is the outcome of a decision-making process which occurred outside the community of national rule-enforcers, then it may encounter more serious problems upon implementation. This casts some doubt on whether it is always appropriate for the IMF to act as a rule-maker—in particular in the aforementioned area of codes of good practices—and, simultaneously, as a standard-enforcer.

A third challenge is that of *developing interagency cooperation*. The traditional barriers in the financial services industry have become increasingly blurred over the past decade. Both the functional and the institutional segmentation of financial markets have eroded, thus making the traditional classifications of financial products and intermediaries obsolete. The need for cooperation has also increased owing to the emergence of financial conglomerates and the increasing importance of universal banking.

At the political level, the *Financial Stability Forum* (FSF) is a relevant achievement in this area. Established in 1999 in response to G7 proposals, the FSF is a composite entity, which is still developing its profile. Its main objectives are to assess the vulnerabilities in the global financial system, to identify and oversee the initiatives needed to address such vulnerabilities, and to improve the coordination between the relevant authorities. The roles played by the FSF tend to range from political impulse to rule-making and rule enforcement, which may not always be advisable, particularly as this could overlap with the work carried out by other fora. The FSF's importance should not, however, be underestimated, as it provides a privileged environment for international interagency cooperation.

At a more technical level, the *Joint Forum of Financial Conglomerates*—consisting of members of the Basel Committee, IOSCO, and the IAIS, as well as the successor of the Tripartite Group—is the main example of interagency cooperation. It has the mandate to improve the exchange of information between banks and securities and insurance supervisors and to develop principles for the future supervision of financial conglomerates. The Forum comprises 12 industrial countries (G10 area and Spain) and has become very productive. In this respect, it is worth mentioning its recent reports on risk concentration principles and intragroup transactions and exposures principles. However, with regard to the supervision of financial conglomerates, the current state of affairs it still far from providing a concrete and harmonized regulatory framework. This is a natural consequence of, on the one hand, the asymmetry in the degree of harmonization of bank regulation and, on the other, the regulation of securities firms and investment companies. In this respect, the Basel Committee's efforts should be supported in order to extend the scope of the revised accord for the purpose of "capturing" risks in the whole banking group, including those bank subsidiaries which are securities and insurance entities.

To conclude with regard to the functioning of international standard-setters, improving the method and logistics of interagency cooperation remains a crucial and challenging unresolved issue. It is indeed very difficult to identify the optimal trade-off between the most effective and the highest possible level of participation in the international cooperation process.

A fourth, and more specific, challenge consists of *closing paradises*, namely tackling the problem of those offshore centers that do not conform to international regulatory and supervisory standards. Such centers can either be "laggards" or behave as "free riders." The first group includes those offshore centers that are willing to cooperate and comply with international standards, but which are characterized by inadequate infrastructures. The second group comprises those jurisdictions which are not willing to cooperate with the international community in a deliberate effort to attract customers by offering them "laxity" (e.g., a lenient fiscal and regulatory legislation). It is not true that these offshore centers are too small to create problems of systemic stability. In an increasingly interdependent M-IMS, it would be shortsighted to ignore the role which free riders play as a shelter for imprudent or illegal activity. The proliferation of competing financial centers, some of which attract business by lowering their prudential standards, has a dynamic component if one considers the number of existing small countries which lack the critical mass to promote sound and efficient financial systems. Of the 200 countries and 6 billion people in the world, one-quarter has a population of less than 1 million, while the ten most populated countries account for around 60 percent of the world population.

The FSF is currently dealing with offshore centers. One of its first achievements was the recent publication of a ranking of centers in three groups on the basis of their willingness to cooperate, the degree of efficiency of their supervisory authorities, and their overall legal infrastructure and adherence to international standards. Another substantial contribution may result from the assessment to be carried out by the IMF in order to enhance offshore centers' application of, and adherence to, international standards. Involvement is also expected from the World Bank and, at the level of industrial countries, the OECD. An important role will continue to be played by the public disclosure of assessments of the progress made by the "critical" offshore centers in adhering to international standards.[9]

5.   Conclusions

The fundamental reason why increased capital mobility constitutes a challenge for the regulation of capital markets is that, while in the Old World the objective of stability was (also) pursued via capital controls by sacrificing the objective of efficiency, in the New World the burden of providing stability has shifted to regulation and supervision while the free movement of capital is allowed to operate as a mechanism for pursuing efficiency.

There has been a change in the roles played by markets and policy-makers in the implementation of the allocation and stabilization functions. In the Old World, we had enough (perhaps even too much) "government" and too little "market." In the New World we have enough (perhaps even too much) "market" and too little "government." The current configuration of the M-IMS, therefore, calls for a strengthening of its institutional framework. It is a long journey, upon which we have only just embarked, and it will take time to reach our goal.

---

[9]   In addition, *direct* disclosure by the offshore centers should also be enhanced. Priority should be given to the implementation of an FSF recommendation according to which the BIS should encourage centers with the highest financial activity to report on financial statistics to be included in the BIS publications.

Appendix

1. International Standard-Setting Fora

	BCBS	IOSCO	IAIS	CPPS	IASC
Area of work	Banks	Securities firms	Insurance companies	Payment and settlement systems	International accounting
Membership or representation	G10 countries plus ECB, European Commission, and FSF with an observer status. Both central banks and supervisory agencies participate in the Committee.	Worldwide (165 organisations representing more than 90 countries). Different levels of membership: ordinary members, affiliate members, and associate members.	Worldwide (insurance supervisors from over 100 jurisdictions). Observers are officially involved in the process of consultation.	Central banks of the G10 countries, Hong Kong SAR and Singapore, as well as the ECB.	Worldwide (143 professional accounting bodies in 104 countries, including 5 associate and 4 affiliate members).
Year of establishment	1974	1983	1994	1990	1973
Secretariat and other resources	Secretariat with large resources, comprising 15 people and located in the BIS (Basle). The BCBS also benefits from contributions by the G10 authorities via subgroups.	General Secretariat located in Madrid.	Secretariat located in the BIS and comprising 5 people.	Secretariat located in the BIS and comprising 6 people. The CPSS also benefits from contributions by the participating central banks via subgroups.	Staff providing technical support, located in London and comprising 8 people.
Frequency of the general meetings of the forum	Regularly, 4 meetings per annum	Annual Conference	Annual Conference	Regularly, 3 meetings per annum	Regularly, 4 meetings per annum
Nature of discussion	Multilateral discussion	Unilateral speeches, which can be held in any language chosen by the host country	Multilateral discussion	Multilateral discussion	Multilateral discussion

2. Main Contributions Released by the International Standard-Setting Fora

1. Basel Committee on Banking Supervision (BCBS)

Banking secrecy and international cooperation in banking supervision (December 1981)

Principles for the supervision of banks' foreign establishments (May 1983)

International convergence of capital measurement and capital standards (July 1988)

Prevention of criminal use of the banking system for the purpose of money laundering (December 1988)

Information flows between banking supervisory authorities (April 1990)

Minimum standards for the supervision of international banking groups and their cross-border establishments (July 1992)

Prudential supervision of banks' derivatives activities (December 1994)

Amendment of the capital accord to incorporate market risks (January 1996)

Core Principles for effective banking supervision (September 1997)

Principles for the management of interest rate risk (September 1997)

Risk management for electronic banking and electronic money activities (March 1998)

Framework for internal control systems in banking organizations (September 1998)

Enhancing bank transparency (September 1998)

Sound practices for banks' interactions with highly leveraged institutions (January 1999)

A new capital adequacy framework (June 1999)

Best practices for credit risk disclosure; Principles for the management of credit risk (July 1999)

Supervisory guidance for managing settlement risk in foreign exchange transactions (July 1999)

Sound Practices for loan accounting and disclosure (July 1999)

Enhancing corporate governance for banking organizations (September 1999)

The Core Principles methodology (October 1999)

Sound practices for managing liquidity in banking organizations (February 2000)

Basel Committee review of international accounting standards (April 2000)

2. International Organization of Securities Commissions (IOSCO)

Capital Adequacy Standards for Securities Firms, Report of the Technical Committee, (October 1989)

Operational and Financial Risk Management Control Mechanisms for Over-the-Counter Derivatives Activities of Regulated Securities Firms, Report of the Technical Committee (July 1994)

Disclosure Framework for Securities Settlement Systems, Joint Report by the Committee on Payment and Settlement Systems and by the Technical Committee (February 1997)

Risk Management and Control Guidance for Securities Firms and their Supervisors, A Report by the Technical Committee (May 1998)

International Disclosure Standards for Cross-Border Offerings and Initial Listings by Foreign Issuers, Report of IOSCO (September 1998)

Objectives and Principles of Securities Regulation (September 1998)

Hedge Funds and other Highly Leveraged Institutions, Report by the Technical Committee of IOSCO (November 1999)

Causes, Effects and Regulatory Implications of Financial and Economic Turbulence in Emerging Markets, Report by the Emerging Markets Committee of IOSCO (November 1999)

3. International Association of Insurance Supervisors (IAIS)

Guidance on Insurance Regulation and Supervision for Emerging Market Economies (September 1997)

Insurance Supervisory Principles (September 1997)

Supervisory Standard on Derivatives (October 1998)

Supervisory Standard on Licensing (October 1998)

Principle Applicable to the Supervision of International Insurers and Insurance Groups and their Cross-Border Business Operations (first version: September 1997; revised version: December 1999)

Principles for the Conduct of Insurance Business (December 1999)

Supervisory Standard on Asset Management by Insurance Companies (December 1999)

4. Committee on Payment and Settlement Systems (CPSS)

Report of the Committee on Interbank Netting Schemes of the Central Banks of the Group of Ten Countries (November 1990)

Central Bank Payment and Settlement Services with Respect to Cross-Border and Multi-Currency Transactions (September 1993)

Disclosure Framework for Securities Settlement Systems (February 1997)

Real-Time Gross Settlement Systems (March 1997)

OTC Derivatives: Settlement Procedures and Counterparty Risk Management (September 1998)

Core Principles for Systemically Important Payment Systems, Consultative document (December 1999)

5. International Accounting Standards Committee (IASC)

Presentation of Financial Statements (reference number of the accounting standard: IAS 1)
Income Taxes (IAS 12)
Information Reflecting the Effects of Changing Prices (IAS 15)
The Effects of Changes in Foreign Exchange Rates (IAS 21)
Borrowing Costs (IAS 23)
Financial Reporting in Hyperinflationary Economies (IAS 29)
Disclosures in the Financial Statements of Banks and Similar Financial Institutions (IAS 30)
Financial Instruments: Disclosures and Presentation (IAS 32)
Financial Instruments: Recognition and Measurement (IAS 39)

6. Joint Forum on Financial Conglomerates

Supervision of Financial Conglomerates (February 1999)
Intra-Group Transactions and Exposures Principles (December 1999)
Risk Concentration Principles (December 1999)

Bibliography

Padoa-Schioppa, T. (1992). *L'Europa verso l'Unione Monetaria*. Torino: Einaudi.
Padoa-Schioppa, T., and F. Saccomanni (1994). Managing a Market-Led Global Financial System. In P.B. Kenen (ed.), *Managing the World Economy: Fifty Years after Bretton-Woods*. Washington, D.C.: Institute for International Economics.
Wallich, H.C. (1972). The Monetary Crisis of 1971: The Lessons to Be Learned. Proceedings of the Lecture Meeting of the Per Jacobsson Foundation, held in Washington.

Sebastian Edwards

Capital Mobility and Economic Performance: Are Emerging Economies Different?

1. Introduction

Political opposition to "globalization" has grown rapidly during the last few years. Protesters in Seattle, Washington, D.C., and other cities around the world have rallied against the alleged evils of an increasingly interconnected world economy, and of the so-called "Washington Consensus."

The opening of domestic capital markets to foreigners is, perhaps, the most reviled aspect of this "consensus." In rejecting a higher degree of capital mobility across countries, the antiglobalization activists and protesters are not alone. Indeed, a number of academics have argued that the free(er) mobility of private capital during the 1990s was behind the succession of crises that the emerging markets experienced during that decade. According to this view, increased capital mobility inflicts many costs and generates (very) limited benefits to the emerging nations. It has been argued that, since emerging markets lack modern financial institutions, they are particularly vulnerable to the volatility of global financial markets. This vulnerability, the story goes, will be higher in countries with a more open capital account. Moreover, many global-skeptics have argued that there is no evidence supporting the view that a higher degree of capital mobility has a positive impact on economic growth in the emerging economies (Rodrik 1998).

Surprisingly, the debate on the effects of capital mobility on economic performance has been characterized by a very limited number of empirical analyses. Some exceptions are Rodrik (1998), Klein and Olivei (1999), Quinn (1997), and Reisen and Soto (2000). The purpose of this paper is to analyze empirically the relationship between economic performance and capital mobility in the world economy. I am particularly interested in understanding two related issues: First,

Remark: I am indebted to Denis Quinn and to Gian Maria Milesi-Ferretti for making their data available, and to Igal Magendzo and Alejandrín Jara for helpful research assistance. I thank Ricardo Hausmann for his comments. I have benefited from discussions with Ed Leamer.

is there any evidence, at the cross-country level, that higher capital mobility is associated (after controlling for other factors) with higher growth? And, second, is the relationship between capital mobility and growth different for emerging and advanced countries?

The paper is organized as follows: Section 1 is the introduction. In Section 2, I provide an analysis of the magnitude, importance, composition, and other characteristics of capital flows in the world economy between 1975 and 1997. In Section 3, I deal with measurement problems. I argue that the complications associated with measuring the actual, as opposed to legal, degree of capital mobility makes the analysis of the connection between capital mobility and growth particularly difficult. In this section, I discuss the properties of various measures on the degree of capital mobility recently constructed by a number of analysts. In Section 4, I report the results from a series of cross-country regressions on economic performance. I focus on two independent variables—GDP growth and total factor productivity growth—and I control for the standard variables, including human capital and the initial degree of economic development. Finally, Section 5 is the conclusions.

2.   Capital Flows in the World Economy during 1975–1997

In this section, I focus on the behavior of capital flows in the world economy during the last two decades. The main objective of this analysis is to unearth regularities, and to detect differences across groups of countries. I consider six groups of countries that correspond to the IMF's *International Financial Statistics* classification: (1) Industrial, (2) African, (3) Asian, (4) Nonindustrial European, (5) Middle East, and (6) Western Hemisphere or Latin America and the Caribbean. In addition, I make a distinction between three types of capital flows: (1) Foreign direct investment (FDI); (2) debt flows, including debt to banks and bonds purchased by foreigners; and (3) other types of flows, mostly portfolio equity flows.

In Tables 1 through 3, I summarize the behavior of capital flows to these six regions and the subtotal "emerging countries" during the period under study. I provide data on averages, medians, standard deviations, and coefficients of variation for the volumes of flows relative to GDP.[1] While Table 1 contains data for the complete period, Tables 2 and 3 present data for each category of capital

---

[1]   These data have been constructed as differences in stocks. The raw data comes from the World Bank.

flow—debt, FDI, equities and total flows—for three different subperiods: The first period is 1975–1982, and corresponds to the years prior to the debt crisis of 1982. The second subperiod is 1983–1989, and corresponds to the years when most emerging countries had difficulties attracting foreign capital. The final subperiod is 1990–1997, a period when private capital flowed, once again, into the emerging economies. This period also corresponds to the initiation of market-oriented reforms in most regions in the world, including the former communist nations.

Visual inspection of Tables 1 through 3 suggests that capital flows have behaved differently across categories, regions, and periods. Flows appear to have been more volatile in the emerging economies and in particular in Africa. These tables also capture the slowdown in flows in the period 1983–1989, when most of the emerging world was battling the consequences of the debt crisis, and the resumption of capital flows in the 1990s, including the surge of portfolio flows. It is also apparent from these figures that Africa has been lagging behind other emerging nations in most capital flow categories.

In order to test formally whether capital flows behaved differently across countries, I estimated a series of nonparametric Kruskal–Wallis χ^2 tests on the equality of the distribution of capital flows in each of the five emerging market regions and the industrial countries. The null hypothesis is that the data from the industrial nations and from each of the emerging regions have been drawn from the same population. The Kruskal–Wallis χ^2 test is computed as

[1] $\qquad K = \{(12/n(n+1)] \Sigma (R_j^2 / n_j)\} - 3(n+1),$

where n_j is the sample size for the j group ($j = 1, ..., m$), n is the sum of the n_js, R_j is the sum of the ranks j group, and the sum Σ runs from $j = 1$ to $j = m$.

The results are reported in Table 4. As may be seen, these tests clearly indicate that capital flows have behaved differently in emerging markets (as a group) and in the industrial countries. With the exception of FDI in the 1975–1982 period, the χ^2 test statistic is larger than the critical value, for every type of flow and for every subperiod considered in this study. The more detailed analysis by region reveals that in most subperiods and for most regions the hypothesis of equality of distribution is rejected strongly. In the case of Africa the χ^2 test statistics are particularly large, and are highly significant in 15 out of 16 cases. During the more recent 1990–1997 period the χ^2 test statistic is below the critical value for FDI flows to Asia, Europe, and Western Hemisphere; it is also below its critical value for equity flows into Europe. In spite of these few instances of χ^2 test statistics below the critical value, the overall picture that emerges from Table 4 indicates, quite strongly, that when it comes to capital flows (relative to GDP) to the emerging markets—both as a broad group and as regional aggregates—they have indeed been different than capital flows to the industrial nations.

Table 1: Capital Flows as Percentage of GDP: Regional Data, 1975–1997

		N	Mean	Median	SD	Coeff. of var.
Debt/GDP	ind	468	3.803	2.542	5.808	1.527
	emerging	2.439	0.763	0.698	15.733	20.615
	afr	1.041	1.564	0.815	15.107	9.658
	asi	411	1.292	0.563	3.662	2.833
	eur	122	1.074	0.678	3.881	3.614
	meast	215	1.478	0.658	5.317	3.597
	westh	650	−1.149	0.737	23.205	−20.194
FDI/GDP	ind	472	0.496	0.168	1.429	2.880
	emerging	1.934	0.225	0.066	2.732	12.140
	afr	797	0.127	0.000	2.823	22.239
	asi	276	0.667	0.066	2.026	3.037
	eur	111	0.497	0.046	1.222	2.461
	meast	160	0.341	0.097	1.656	4.854
	westh	590	0.068	0.221	3.263	47.690
Equity/GDP	ind	472	0.238	0.000	1.137	4.767
	emerging	2.042	0.022	0.000	0.354	16.271
	afr	868	0.011	0.000	0.212	18.914
	asi	316	0.099	0.000	0.434	4.399
	eur	111	0.052	0.000	0.195	3.732
	meast	179	−0.103	0.000	0.875	−8.511
	westh	568	0.028	0.000	0.143	5.046
Total/GDP	ind	468	4.529	3.116	6.334	1.398
	emerging	1.853	1.683	1.086	12.094	7.185
	afr	744	1.838	0.921	18.002	9.793
	asi	272	2.396	0.986	5.250	2.191
	eur	111	1.482	1.248	3.537	2.386
	meast	158	1.713	1.257	4.346	2.537
	westh	568	1.170	1.197	5.626	4.810

Source: World Bank, WDI CD-ROM, 1999.

Table 2: Debt and FDI Capital Flows as Percentage of GDP: Regional Data and Alternative Periods, 1975–1997

		N	Mean	Median	SD	Coeff. of var.	N	Mean	Median	SD	Coeff. of var.
				Debt					*FDI*		
1975–1982	ind	158	3.328	2.165	4.550	1.367	160	0.099	0.026	1.063	10.763
	emerging	827	1.116	1.477	21.504	19.273	669	0.146	0.011	2.374	16.207
	afr	357	2.725	1.698	25.114	9.215	288	0.131	0.000	2.916	22.185
	asi	127	1.707	0.700	2.962	1.735	87	0.492	0.002	1.605	3.260
	eur	37	2.260	1.001	5.895	2.608	32	0.053	0.000	0.937	17.756
	meast	80	2.341	1.593	4.074	1.740	59	0.701	0.157	1.766	2.519
	westh	226	-2.380	1.682	25.808	-10.844	203	-0.127	0.083	2.053	-16.183
1983–1989	ind	146	3.926	2.891	3.944	1.005	147	0.496	0.136	1.389	2.800
	emerging	764	1.019	0.743	10.281	10.088	621	-0.062	0.000	3.183	-51.691
	afr	324	1.990	1.347	5.131	2.578	259	-0.054	-0.026	2.093	-38.747
	asi	132	1.020	0.656	2.820	2.764	90	0.456	0.003	1.974	4.330
	eur	36	0.338	0.323	2.545	7.531	35	0.206	0.000	0.769	3.739
	meast	68	1.402	0.732	3.075	2.193	49	0.422	0.058	1.533	3.632
	westh	204	-0.531	0.556	18.492	-34.811	188	-0.496	0.000	4.959	-10.007
1990–1997	ind	164	4.151	2.508	7.907	1.905	165	0.882	0.545	1.658	1.881
	emerging	848	0.189	0.196	12.877	68.222	644	0.583	0.221	2.569	4.406
	afr	360	0.029	0.134	2.860	97.275	250	0.309	0.017	3.327	10.759
	asi	152	1.182	0.422	4.687	3.964	99	1.012	0.252	2.350	2.323
	eur	49	0.719	1.101	2.358	3.280	44	1.051	0.582	1.480	1.408
	meast	67	0.525	-0.101	7.771	14.797	52	-0.144	0.086	1.546	-10.740
	westh	220	-0.458	0.114	24.333	-53.178	199	0.800	0.582	1.808	2.258

Source: World Bank, WDI CD-ROM, 1999.

Table 3: Portfolio and Total Capital Flows as Percentage of GDP: Regional Data and Alternative Periods, 1975–1997

		N	Mean	Median	SD	Coeff. of var.	N	Mean	Median	SD	Coeff. of var.
				Portfolio					*Total Capital*		
1975–1982	ind	160	-0.059	0.000	0.380	-6.426	158	3.352	2.385	4.871	1.453
	emerging	717	0.022	0.000	0.369	16.794	661	2.573	1.876	18.970	7.374
	afr	314	0.007	0.000	0.206	27.603	280	2.976	1.640	28.677	9.637
	asi	101	0.057	0.000	0.230	4.046	87	2.601	1.258	4.274	1.643
	eur	32	0.000	0.000	0.000		32	2.023	1.233	4.864	2.404
	meast	67	0.115	0.000	1.086	9.461	59	3.129	2.702	4.278	1.367
	westh	203	0.000	0.000	0.021	-128.270	203	1.930	1.976	4.751	2.462
1983–1989	ind	147	0.38	0.000	1.571	4.947	146	4.756	3.463	5.171	1.087
	emerging	680	-0.021	0.000	0.295	-14.163	610	1.470	0.726	5.451	3.709
	afr	301	-0.005	0.000	0.095	-18.461	252	1843	0.903	5.553	3.013
	asi	104	0.048	0.000	0.303	6.337	90	1.616	0.727	4.112	2.544
	eur	35	0.000	0.000	0.001	5.916	35	0.520	0.274	2.791	5.371
	meast	56	-0.320	0.000	0.866	-2.703	49	1.990	1.109	3.820	1.920
	westh	184	0.002	0.000	0.014	7.810	184	0.929	0.436	6.508	7.004
1990–1997	ind	165	0.457	0.175	1.111	2.432	164	5.461	3.957	8.126	1.488
	emerging	645	0.066	0.000	0.386	5.822	582	0.897	0.645	4.967	5.540
	afr	253	0.035	0.000	0.302	8.528	212	0.330	0.351	3.624	10.969
	asi	111	0.184	0.000	0.627	3.406	95	2.947	1.033	6.768	2.297
	eur	44	0.131	0.049	0.294	2.236	44	1.855	1.700	2.772	1.495
	meast	56	-0.145	0.000	0.450	-3.097	50	-0.229	-0.417	4.268	-18.635
	westh	181	0.087	0.000	0.241	2.772	181	0.562	0.667	5.499	9.789

Source: World Bank, WDI CD-ROM, 1999.

Table 4: Kruskal–Wallis Test for Equality of Samples[a]

		emerg	afr	asi	eur	meast	westh
1975–1997	Total/GDP	127.5901	90.05093	64.26101	35.84606	31.06249	96.60054
	Debt/GDP	130.8848	84.52159	99.65771	44.21965	49.21661	98.44996
	FDI/GDP	7.687368	23.13025	0.012958	0.130548	0.815106	1.280828
	Equity/GDP	15.74963	19.95323	0.883731	0.36937	13.79062	8.802971
1975–1982	Total/GDP	3.618905	3.045184	4.340916	4.578451	0.09373	1.68014
	Debt/GDP	9.003849	3.964778	16.46382	8.239757	3.055167	4.076299
	FDI/GDP	0.935678	0.12325	2.07707	0.01943	6.213636	0.104623
	Equity/GDP	8.022592	4.419214	9.852856	1.644552	3.45189	4.781673
1983–1989	Total/GDP	71.24705	39.14418	48.53438	27.80614	13.28654	62.84356
	Debt/GDP	58.00618	24.95593	58.57322	26.92267	23.88016	51.57387
	FDI/GDP	10.43872	14.6369	1.418094	1.143263	0.573863	6.81855
	Equity/GDP	25.20113	23.39485	6.571476	5.244911	14.68386	15.7491
1990–1997	Total/GDP	78.70559	69.62941	22.20793	11.38153	34.12221	52.93491
	Debt/GDP	68.82774	62.18753	27.93487	12.8117	27.22733	51.25942
	FDI/GDP	7.20749	25.67238	0.380532	1.197198	11.58874	0.037483
	Equity/GDP	24.89199	25.9106	5.946005	1.955913	18.64987	13.11557

[a]The null hypothesis is that each emerging region comes from the same population as the industrial countries. The test is distributed χ^2 with one degree of freedom. The critical value at the 10 percent level is 2.71.

3.   Measuring the Extent of Capital Mobility in Emerging and Advanced Economies

During most of the last 50 years the vast majority of what we today call emerging nations severely controlled international capital movements. This was done through a variety of means, including taxes, administrative restrictions, and outright prohibitions. It has only been in the last decade or so that serious consideration has been given to the opening of the capital account in less advanced nations. Many analysts have associated the proposals for free capital mobility with the policy dictates of the so-called "Washington Consensus." Williamson's (1993) original article on the Washington Consensus, however, says very little about the opening of the capital account. What it does say, however, is that the reform policies favored by the multilaterals included encouraging foreign direct investment and the liberalization of domestic capital markets.

Legally speaking—and as the IMF documented year after year—during most of the post World War II era, the vast majority of the emerging countries had a closed capital account. From an economic point of view, however, what matters is not the *legal* degree of capital restrictions, but the actual or "true" degree of capital mobility. There is ample historical evidence suggesting that there have been significant discrepancies between the legal and the actual degree of capital controls. In countries with severe impediments to capital mobility, including countries that have banned capital movement, it does not take a long time for the private sector to find ways to get around the restrictions. The most common mechanisms have been the overinvoicing of imports and underinvoicing of exports. The massive volumes of capital flight that took place in Latin America in the wake of the 1982 debt crisis clearly showed that, when faced with the "appropriate" incentives, the public can be extremely creative in finding ways to move capital internationally.

a.   Previous Measurement Attempts

Measuring the "true" degree of capital mobility is not easy, and is still subject to considerable debate. In an early study, Harberger (1980) argued that the effective degree of integration of capital markets should be measured by the convergence of private rates of return to capital across countries. He used national accounts data for a number of countries, including eleven Latin American countries, to estimate rates of return to private capital, and found out that these were significantly similar. More importantly, he found that these private rates of return were independent of national capital-labor ratios. Harberger interpreted these findings as supporting the view that capital markets are significantly more integrated than what a simple analysis of legal restrictions would suggest. In an effort to measure the "true" degree of capital mobility, Feldstein and Horioka (1980) analyzed the behavior of savings and investments in a number of countries. They argued that if there is perfect capital mobility, changes in savings and investments will be uncorrelated in a specific country. Using a data set for 16 OECD countries they found that savings and investment ratios were highly positively correlated, and concluded that these results strongly supported the presumption that *long-term* capital was subject to significant impediments. Frankel (1989) applied the Feldstein–Horioka test to a large number of countries during the 1980s, including a number of Latin American nations. His results corroborated those obtained by the original study, indicating that savings and investment were significantly positively correlated in most countries. Montiel (1994) estimated a series of Feldstein–Horioka equations for emerging countries. He argues that the estimated regression coefficient for the industrial countries could be used as a benchmark for evaluating whether a particular country's capital account is open or not. After analyzing a number of studies, he concludes that a saving ratio regression coefficient of 0.6 provides an adequate benchmark: if a country's regression coefficient exceeds 0.6, it can be classified as having a "closed" capital account; if the coefficient is lower than 0.6, the country has a rather high degree of capital mobility. Using this procedure, he concludes that many emerging nations have exhibited a remarkable degree of capital mobility—that is, much larger than what an analysis of legal restrictions would suggest.

In a series of studies, Edwards (1985, 1988) and Edwards and Khan (1985) argued that time series on domestic and international interest rates could be used to assess the degree of openness of the capital account (see also Montiel 1994). The application of this model to the cases of a number of countries (Brazil, Colombia, Chile) confirmed the results that, in general, the actual degree of capital mobility is greater than what the legal restrictions approach suggests. Haque and Montiel (1991), Reisen and Yeches (1993), and Dooley (1995) have provided expansions of this model that allow for the estimation of the degree of capital mobility even in cases when there are not enough data on domestic interest rates, and when there are changes in the degree of capital mobility through time. Their results once again indicate that in most Latin American countries "true" capital mobility has historically exceeded the "legal" extent of capital mobility. Dooley et al. (1997) have developed a model that recognizes the costs of undertaking disguised capital inflows. The model is estimated using a Kalman filter technique for three countries, including Mexico. The results suggest that Mexico, the Philippines, and Korea experienced a very significant increase in the degree of capital mobility between 1977 and 1989.

More recently, some authors have used information contained in the International Monetary Fund's *Exchange Rate and Monetary Arrangements* to construct indexes on capital controls for panels of countries. Alesina et al. (1994), for example, constructed a dummy variable index of capital controls. This indicator—which takes a value of one when, according to the IMF, capital controls are in place and zero otherwise—was then used to analyze some of the political forces behind the imposition of capital restrictions in a score of countries. Rodrik (1998) used a similar index to investigate the effects of capital controls on growth, inflation, and investment between 1979 and 1989. His results suggest that, after controlling for other variables, capital restrictions have no significant effects on macroeconomic performance. Klein and Olivei (1999) used the IMF's *Exchange Rate and Monetary Arrangements* data to construct an index of capital mobility. The index is defined as the number of years in the period 1986–1995 that, according to the IMF, the country in question has had an open capital account.[2] In contrast to Rodrik, their analysis suggests that countries with a more open capital account have performed better than those that restrict capital mobility.

An important limitation of these IMF-based indexes, however, is that they are extremely general and do not distinguish between different intensities of capital restrictions. Moreover, they fail to distinguish between the type of flow that is being restricted, and they ignore the fact that, as discussed above, legal restrictions are frequently circumvented. For example, according to this IMF-based indicator, Chile, Mexico, and Brazil were subject to the same degree of capital controls in 1992–1994. In reality, however, the three cases were extremely different. While in Chile there were restrictions on short-term inflows, Mexico had (for all practical purposes) free capital mobility, and Brazil had in place an arcane array of restrictions. Montiel and Reinhart (1999) have combined IMF and country-specific information to construct an index on the intensity of capital controls in 15 countries during 1990–1996. Although their index, which can take three values (0, 1, or 2), represents an improvement over straight IMF indicators, it is still very general, and does not capture the subtleties of actual capital restrictions.

Quinn (1997) has constructed the most comprehensive set of cross-country indicators of the degree of capital mobility. His indicators cover 20 advanced countries and 45 emerging economies. These indicators have two distinct advantages over other indicators: First, they are not restricted to a binary classification, where countries capital account's are either open or closed. Quinn uses a 0 through 4 scale to classify the countries in his sample, with a higher number meaning a more open capital account. Second, Quinn's indicators cover more than one time period, allowing researchers to investigate whether there is a connection between capital account *liberalization* and economic performance. This is, indeed, a significant improvement over traditional indexes that have concentrated on a particular period in time, without allowing researchers to analyze whether countries that open up to international capital movements have experienced changes in performance.

b.   A Comparison of Two Alternative Measures of the Extent of Capital Mobility

In this subsection, I analyze the main properties of two broad measures of capital mobility:[3] (1) An index based on the number of years within a certain period that,

---

2   Milner (1996) and Razin and Rose (1994) have also used indicators based on the IMF binary classification of openness.

3   I am grateful to Gian Maria Milesi-Ferretti for making his data set available to me.

according to the IMF, a particular country has not imposed capital controls. This is the type of indicator used by Alesina et al. (1994), and by Rodrik (1998) in his study on the relationship between capital controls and economic performance. I call this index NUYCO. A higher value denotes a higher degree of *capital controls*. And (2) Quinn's (1997) index of capital mobility. This indicator can take values from 0 through 4, with increments of 0.5. A higher value of this index denotes a higher degree of *capital mobility*.

The *number-of-years-with-controls (NUYCO) index:* I computed this index for three periods, 1981–1985, 1986–1990 and 1981–1990, for a sample of 61 countries. Of these, 40 are emerging nations and 21 are advanced countries. Panel A in Table 5 contains summary statistics. Two properties of this index emerge from this table. First, as expected, capital controls have been more pervasive among the emerging countries. Second, while the advanced countries appear to have relaxed capital controls somewhat during the second half of the 1980s, the emerging countries appear to have tightened them slightly. In order to analyze formally whether this index is statistically different in industrial and emerging countries, I computed, once again, Kruskal–Wallis test statistics. The χ^2 was (p value in parentheses): 2.33 (0.12) for 1981–1985, 5.57 (0.018) for 1985–1990, and 3.67 (0.057) for 1981–1990. Overall these test statistics confirm the hypothesis that the extent of capital controls was significantly larger in the emerging countries, especially in the second half of the 1990s.[4]

One of the most serious limitations of the NUYCO index is that it tends to classify most countries in the extremes, as either being subject to no controls or as completely impeding capital mobility. For the period 1981–1985, only 8 out of 61 countries have values different from 0 or 5, and for 1985–1990 only 3 countries have an index value different from the extremes of 0 or 5. As pointed out above, this inability to consider intermediate cases of limited controls is one of the greatest shortcomings of this index. In my view, this problem is so severe that it reduces very significantly its usefulness in empirical cross-section analyses.[5]

Quinn's Indicator (CAPOP): Panel B in Table 5 contains summary statistics for Quinn's index of capital account restrictions. As was pointed out, this index can take values that go from 0 to 4, and is available for 65 countries—20 of which are industrial and 45 emerging—and for two periods: the mid-1970s and

---

[4]   This could be, in part, a result of these countries' efforts to get over the debt crisis of 1982. Formally testing this proposition, however, is beyond the scope of this paper.

[5]   The index used by Klein and Olivei (1999) also suffers from this limitation. The very vast majority of countries in their sample appear to be either completely open or completely closed to capital mobility.

Table 5: Alternative Indicators of Capital Mobility: Summary Statistics

A. Nuyco Index

	Obs.	Mean	Std. dev.	Min.	Max.
	All countries				
NUYCO85	61	3.737705	2.040438	0	5
NUYCO90	61	3.737705	2.136209	0	5
NUYCO819	61	7.47541	4.002527	0	10
	Industrial countries				
NUYCO85	21	3.142857	2.329929	0	5
NUYCO90	21	2.714286	2.452404	0	5
NUYCO819	21	5.857143	4.607447	0	10
	Emerging countries				
NUYCO85	40	4.05	1.82504	0	5
NUYCO90	40	4.275	1.753933	0	5
NUYCO819	40	8.325	3.407289	0	10

B. Quinn Index (CAPOP)

Variable	Obs.	Mean	Std. dev.	Min.	Max.
	All countries				
CAPOP73	65	2.069231	1.089283	0	4
CAPOP87	65	2.338462	1.142597	0	4
	Industrial countries				
CAPOP73	21	2.452381	.756716	1.5	4
CAPOP 87	21	3.333333	.5773503	2.5	4
	Emerging markets				
CAPOP 73	44	1.886364	1.180577	0	4
CAPOP 87	44	1.863636	1.036338	0	4

Source: Alesina et al. (1994), Rodrik (1998), and Quinn (1997).

the mid/late 1980s.[6] As may be seen from Table 5, according to this indicator, the advanced countries had a more open capital account than the emerging econ-

---

[6]   The indexes were formally computed for 1973 and 1987. See Quinn (1997) for details.

omies. These data also suggest that the difference between the two groups of countries became more accentuated in the late 1980s. The Kruskal–Wallis χ^2 for equality of the distribution across the two groups of countries is (p value in parentheses): 5.1 (0.02) for the mid-1970s and 23.7 (0.0001) for the 1980s. In contrast with the NUYCO index, Quinn's indicator allows for considerable gradation in the extent of capital mobility. For instance, for the 1970s in 49 of the 65 countries, the index has a value other than the extremes of 0 or 4. In the 1980s, there are 46 countries with an intermediate value for the index—that is a value that is neither 0 (the capital account is completely closed) nor 4 (the capital account is completely open). For example, while, according to NUYCO, Greece and Ireland had a completely closed capital account during the second half of the 1980s; according to Quinn's index, they had a semi-open capital account during this period.

An important feature of Quinn's index is that it has been computed for two different periods, allowing us to investigate the effects of capital account liberalization on economic performance. In order to understand further the properties of Quinn's indexes, I compared them with an indicator based on Montiel's (1994) savings-investments regressions for emerging countries. The Spearman rank correlation coefficient had the expected sign, but was rather low. Moreover, the null hypothesis of both indexes being independent could not be rejected at conventional values.

4.   Capital Mobility and Economic Performance: New Results

In principle, a greater degree of openness of the capital account can impact on economic performance through two alternative channels. The first, and most obvious one, is through its effect on foreign savings, and through them, on aggregate investment. Countries with a more open capital account will have, in principle, the ability to finance a larger current account deficit, and thus increase the volume of foreign savings. If increases in foreign services are not reflected in a one-to-one decline in domestic savings, aggregate savings will be higher. This will allow for higher investment and, thus, faster growth.

Figure 1 presents the relationship between the degree of capital account liberalization during the 1980s—denoted as ($q$open87–$q$open73), and measured in the horizontal axis—and the change in aggregate capital inflows, measured in the vertical axis. This figure suggests, quite strongly, that countries that reduced the degree of capital controls experienced an increase in capital inflows. This, in turn, was translated into higher current account deficits. Whether this, in turn, re-

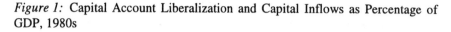

Figure 1: Capital Account Liberalization and Capital Inflows as Percentage of GDP, 1980s

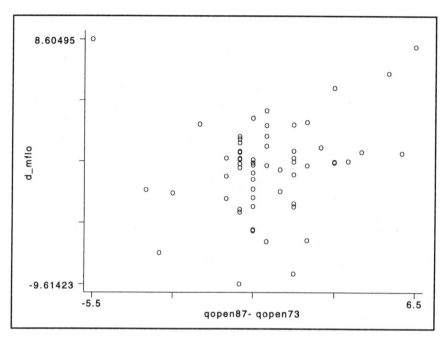

sulted in higher aggregate investment depends on the extent to which foreign savings crowded out domestic savings and is, ultimately, an empirical issue.

In Edwards (1996), I used a broad cross-country data set to analyze this issue. My results suggested that an increase in the current account deficit—that is, an increase in foreign savings—crowded out private domestic savings partially. The point estimate ranged (in absolute value) from 0.38 to 0.625, depending on the specification used for the regression. These results were confirmed by the direct estimation of investment equations that included the CAPOP index of capital mobility as a regressor. These regressions are similar to those estimated by Barro and Sala-i-Martin (1995), and are not reported here due to space considerations.[7]

The second (potential) channel through which capital mobility may affect performance refers to efficiency and productivity growth. According to a number of authors, the elimination of capital controls reduces an important distortion and

---

[7]   Results are available from the author on request.

will tend to result in higher return to investment, and higher productivity growth. That is, according to this channel, countries with more open capital accounts will outperform those with restrictions on capital mobility, even after controlling for the direct investment effect. In this section, I use the data described above to investigate the importance of this particular channel.

a.   Basic Econometric Results

According to economic theory, countries with fewer distortions will tend to perform better than countries with regulations and distortions that impede the functioning of markets. For some time now, most (but not all) economists have agreed that freer trade in goods and services indeed results in faster growth (Barro and Sala-i-Martin 1995, Edwards 1998c). In standard models, this "free trade" principle extends to the case of trade in securities, and countries that have fewer restrictions on capital mobility will, with other things given, tend to outperform countries that isolate themselves from global financial markets. This view is clearly expressed by Rogoff (1999: 23):

> From a theoretical perspective, there are strong analogies between gains in intertemporal trade in goods, and standard intratemporal trade.... In theory, huge long-run efficiency gains can be reaped by allowing global investment to flow towards countries with low capital-labor ratios.... [R]esearchers have now come to believe that the marginal gains of [international] trade in equities can be very large.... [It allows countries] to diversify production risk, which allows smaller countries to specialize, and more generally to shift production towards higher-risk, higher return projects.

Whether gains from an open capital account are as large as Rogoff believes is largely an empirical question. In this section, I use a new cross-country data set to investigate this issue. More specifically, I concentrate on two measures of performance: real GDP growth and total factor productivity growth during the 1980s. I rely on Quinn's index to measure the degree of capital mobility in different countries. The data on GDP growth are taken from Summers and Heston and those on TFP growth are from Edwards (1998c).

From a policy perspective, analysts are interested in two related issues: (1) Have countries with a more open capital account performed better—in terms of higher productivity growth and per capita GDP growth—than countries that restrict capital mobility? (2) Have countries that have opened their capital account performed differently than countries that have not done so? As noted, a particularly important question is whether there is a "performance effect" over and above the investment effect discussed above. One of the advantages of

Quinn's index is that it is available at two different periods in time, allowing us to address both of these questions. The analysis presented in this section investigates the relationship between capital account restrictions and economic performance which is based on the estimation of the following two equations:

[2] $g_j = \alpha_0 + \alpha_1 \kappa_j + \Sigma \alpha_2 X_j + \varepsilon_j$

[3] $\tau_j = \beta_0 + \beta_1 \kappa_j + \Sigma \beta_2 X_j + \mu_j$,

where g_j is average real GDP growth in country j during the 1980s, τ_j is the average rate of TFP growth during the 1980s, κ_j is a measure of capital account openness in country j, or an indicator of the extent of capital account liberalization between 1973 and 1987. The x_js are other variables that affect economic performance; ε_j and μ_j are heteroskedastic errors with zero mean. The αs and βs are parameters to be estimated. Following the recent literature on growth and cross-country economic performance in the estimation of equation [3] the following X_js were included: (a) The investment ratio during the 1980s (INV80). Its coefficient is expected to be positive. (b) A measure of human capital, taken to be the number of years of schooling completed by 1965 (human). Its coefficient is expected to be positive. And (c) the log of real GDP per capita in 1965, which is taken to be a measure of initial economic activity. To the extent that countries' real income tends to converge, the coefficient of this variable (GDP65l) is expected to be negative. In the estimation of equation [3] initial GDP and human capital were used as the two X_js.

The first measure of capital account openness (CAPOP) captures the *degree of openness of the capital account in each country during the mid/late 1980s* and corresponds to Quinn's index discussed above. The second index captures the extent to which capital account restrictions changed between the mid-1970s and the mid/late 1980s. This index, which is denoted D_CAPOP, corresponds to Quinn's capital account liberalization indicator; a higher value means that the country in question liberalized its capital account during the period under study. The sign and statistical significance of the capital account openness coefficient is at the heart of recent discussions on the effects of globalization, and is the main interest of the econometric analysis reported in this section.

Equations [2] and [3] were estimated using a number of procedures, including weighted least squares, weighted two-stage least squares, SURE, and weighted three-stage least squares. In all regressions, GDP per capita in 1985 was used as a weight.

Table 6 summarizes the basic results obtained from the estimation of equations [2] and [3] using the level of capital account restrictions (CAPOP) as the independent variable. In Table 7, on the other hand, I present the regression re-

Table 6: Capital Account Openness and Growth of GDP and Total Factor Productivity (TFP) in the 1980s for All Countries: Cross-Country Econometric Results (weighted least squares (WLS) and instrumental variables (IV-WLS))[a]

	Coef.	Std. err.	t	P>t
Real GDP growth (WLS)				
Capop87	.0030046	.0023936	1.255	0.215
human65	.0031461	.001128	2.789	0.007
gdp65l	−.0123794	.0041281	−2.999	0.004
inv80s	.1603782	.0283357	5.660	0.000
_cons	.0539573	.0283515	1.903	0.062
N = 59;		F(4, 54) =13.14;		R-squared = 0.4932
Real GDP growth (IV-WLS)[a]				
Capop87	.0079586	.0035431	2.246	0.029
human65	.0029805	.001113	2.678	0.010
gdp65l	−.0162367	.0044724	−3.630	0.001
inv80s	.1475412	.0287109	5.139	0.000
_cons	.0765549	.0298211	2.567	0.013
N = 56;		F(4, 51) = 13.70;		R-squared = 0.5023
TFP growth (WLS)				
Capop87	.0053182	.00277	1.920	0.060
human65	.0048019	.0012986	3.698	0.000
gdp65l	−.0124138	.0047	−2.641	0.011
_cons	.0612132	.0329048	1.860	0.068
N = 62;		F(3, 58) = 8.31;		R-squared = 0.3006
TFP growth (IV-WLS)[b]				
Capop87	.0059084	.002569	2.300	0.025
human65	.0019274	.0008243	2.338	0.023
gdp65l	−.0058782	.0033151	−1.773	0.082
_cons	.0223116	.0217966	1.024	0.311
N = 58;		F(3, 54) = 6.84;		R-squared = 0.1758

[a]Instruments: human6, gdp65, qcap7, lly70, inv80, open80, dist, lly75, bmp75l. —
[b]Instruments: human6, gdp65, qcap7, lly70, open80, dist, lly75, bmp75l.

Table 7: Capital Account Liberalization and Growth of GDP and Total Factor Productivity (TFP) in the 1980s for All Countries: Cross-Country Econometric Results (weighted least squares (WLS) and instrumental variables (IV-WLS))[a]

	Coef.	Std. err.	t	P>t
GDP growth (WLS)				
human65	.0028349	.0011286	2.512	0.015
gdp65l	−.0101456	.0038152	−2.659	0.010
inv80s	.1550862	.0281895	5.502	0.000
D_capop	.003052	.0015201	1.627	0.110
_cons	.0443192	.0272326	1.458	0.151
N = 59; F(4; 53) = 14.17; Prob > F = 0.0000; R-squared = 0.5167				
GDP growth (IV-WLS)[a]				
D_capop	.0051239	.0026291	1.949	0.057
gdp65l	−.0096943	.0039819	−2.435	0.019
inv80s	.1515098	.030562	4.957	0.000
human65	.0024621	.0012104	2.034	0.047
_cons	.0416354	.0285659	1.458	0.151
N = 55; F(4, 50) = 12.49; Prob > F = 0.000; R-squared = 0.4883				
TFP growth (V-WLS)[b]				
qdcap738	.0053401	.0016531	3.230	0.002
human65	.00144	.0008524	1.689	0.097
gdp65l	−.0019287	.0027714	−0.696	0.490
_cons	.0042915	.0200442	0.214	0.831
N = 57; F(3, 53) = 8.49; R-squared = 0.1884				
TFP growth (IV-WLS)				
qdcap738	.0025068	.0009653	2.597	0.012
human65	.001953	.000744	2.625	0.011
gdp65l	−.0017833	.0024749	−0.721	0.474
_cons	.0024587	.0178655	0.138	0.891
N = 60; F(3, 56) = 9.24; R-squared = 0.3311				

[a]The following instruments were used: (human65, gdp65l, inv80, qcap58, qopen58, qopen73, qcap73, lly75, dist). — [b]Instruments: (human6, qcap7, lly70, open80, dist, lly75, bmp751).

sults from the estimation of these two equations using the capital account liberalization index, D_CAPOP, as the independent variable.[8] In both tables the sample includes all countries for which data are available.

As may be seen, the estimated coefficients of *human65*, *inv80*, *gdp651* have the expected signs in every regression. Moreover, in the vast majority of the regressions the estimated coefficients for these variables were significant at conventional levels. More important for the subject matter of this paper is that the coefficients of the capital account openness variables are positive in every regression, and significant in all but one of them. These results suggest that, once other variables have been controlled for, countries that are more integrated into global financial markets have performed better than countries that have isolated themselves. This is the case both for countries that had a more open capital account and for countries that liberalized their capital account.

It is interesting to note that if instead of using Quinn's indexes of capital account restrictions, the more common IMF-based indicator is used, the coefficients become insignificant. For instance, in the WLS estimation of the growth equation, the coefficient of NUYCO is 0.0002 with a t-statistic of 0.657. When this equation was estimated using IV-WLS, the coefficient was 0.0008 and the t-statistic 1.12.

b.   Outliers and Measurement Errors

In order to investigate the robustness of these results, I performed a sensitivity analysis: I checked for the possible undue influence of outliers, and I dealt with measurement error.

Outliers: In order to investigate the possible undue influence of outlier observations, I computed Cook's influence distance test. The results point towards three potential outliers: Nicaragua, India and Ethiopia. When these countries were excluded from the estimation, however, the results did not change in any relevant way: countries with a more open capital account and countries that have liberalized capital flows appear to have outperformed more isolationist nations.

Errors in Variables: Even though Quinn's indicators of capital account restrictions are vastly superior than the more traditional ones—including the IMF-based indexes used by Rodrik (1998) and others—they are still an imperfect measure of the "true" degree of capital mobility. In that sense, the estimation of equations [2] and [3] are subject to a classical error-in-variables problem. The

---

[8]   The basic SURE and three-stage least square results are not reported due to space considerations. See, however, the discussion below.

traditional, textbook solution to this problem is to estimate the equation in question using instrumental variables. If the "mismeasured" variable is properly instrumented, the estimated coefficient is consistent. In that sense, then, it is possible to argue that, since the results reported above were obtained with instrumental variables, the measurement problem has been properly tackled.

In the current case, however, the extent of measurement error is likely to be more severe than the simple textbook case. Indeed, in this case all independent variables are (possibly) measured with error. Klepper and Leamer (1984) have shown that when measurement error is generalized, it is possible to use a set of reversed regressions to compute bounds for the coefficients of interest. These authors show that if there are no changes in the pattern of coefficient signs when estimating the reversed regressions, the "true" value of each coefficient will be bounded by the minimum and maximum estimates from the set of reversed and direct regressions. If, on the other hand, there is a change in the sign pattern of *any of the coefficients*, it is necessary to bring in additional information to be able to bound the coefficients.

Following Klepper and Leamer (1984), I estimated the reversed regressions corresponding to equations [2] and [3], and analyzed the sign pattern of the coefficients. Unfortunately, in each equation there were two sign changes, indicating that it is not straightforward to bound the "true" coefficients. This suggests that the estimates reported above are (somewhat) fragile. In order to address this issue further, I estimated the critical minimal level for the R^2 between the dependent and the "true" (error free) explanatory variable that is consistent with a coefficients vector bounded by the original orthant. For the TFP equation on capital account liberalization, this minimum value, or R_m^2, is 0.57. This critical value is not completely unreasonable, indicating that, although the estimated equations [2] and [3] are fragile—in the sense that the reversed regression coefficients exhibit a sign switch—it is possible to assume that their "true" values correspond to those obtained in the direct regressions. In that sense, then, this analysis provides some further support for the finding that, at least for the period under consideration, countries with a greater degree of integration with the rest of the world performed better than more isolated nations.

c.   How Different Are Emerging Countries?

As pointed out in Section 1 of this paper, one of the most important policy questions—and one that is at the heart of recent debates on globalization—is whether the effects of globalization on economic performance is similar in advanced and in emerging economies. In fact, according to many intellectually prominent global skeptics, capital account liberalization is not bad per se. The problem, in their view, is that the emerging countries are unprepared for it. The problem is, according to this view, that the poor nations do not have the required institutions to handle efficiently large movements of capital. In this subsection I provide some preliminary results that address this issue. Due to space considerations I concentrate on the case when the degree of capital account openness is used as the independent variable. More specifically, I investigate whether the effect of capital restrictions on growth depends on the country's level of development. I do this by adding the interactive independent variable (log $GDPC \cdot CAPOP$) in the estimation of equations [2] and [3]. GDPC is GDP per capita in 1980. In this case, equation [2] becomes:

$$[2'] \qquad g_j = \alpha_0 + \alpha_1 \, CAPOP_j + \alpha_2 \, (CAPOP_j \log GDPC_j) + \alpha_3 human65_j +$$

$$+ \, \alpha_{14} \log GDPC65_j + \varepsilon_j.$$

If coefficient α_2 is significant, the total effect of capital openness on growth becomes country-specific, and will be given by:

$$[3] \qquad E_j = \alpha_1 + \alpha_2 \log GDPC_j.$$

If α_2 is positive (negative), the effect of capital account openness on growth increases (declines) with the level of development. Table 8 contains the results obtained from the three-stage least squares estimation of equations [2] and [3], with an added interactive regressor. As may be seen, all the coefficients are significant at conventional levels. More important, however, the coefficient of the interactive term ($CAPOP_j \log GDPC_j$) is positive, indicating that the effect of a more open capital account increases with the initial level of development of the country. Furthermore, since the coefficient of the openness index is *negative*, an open capital account may in fact have a negative effect at very low levels of development.

A particularly important question is what is the average value of the total effect coefficients ($\alpha_1 + \alpha_2 \log GDPC_j$) for industrial and emerging countries? In both equations, these averages were positive for the advanced nations and negative for the emerging countries. The actual averages in the growth equation were (standard errors in parentheses), 0.0075932 (0.0035522) for industrial countries and −0.0104921 (0.0091007) for emerging nations. In the TFP growth equation the industrial average was −0.0133434 (0.0083494), while the emerging markets average was −0.0133434 (0.0083494). Notice, however, that not every emerging country has a negative ($\alpha_1 + \alpha_2 \log GDPC_j$) coefficient. In fact, in the case of the growth equation, the following emerging countries had a positive "total effect" coefficient: Israel, Venezuela, Hongkong, Singapore, Mexico.

Table 8: Capital Account Openness and Growth: The Role of Interactive Terms (weighted three-stage least squares)

Equation	Obs	Parms	RMSE	R^2	χ^2	P
gro80s	56	5	.0117901	0.4990	72.73749	0.000
tfp80	56	4	.0092844	0.0199	23.37243	0.0001
	Coef.	Std. err.	z	p > |z|	[95% conf. interval]	
			GDP growth (gro80s)			
qcap87	−.1070224	.0455303	−2.351	0.019	−.1962602	−.0177846
log_qc8	.0123887	.0050162	2.470	0.014	.0025571	.0222202
human65	.0025725	.0010615	2.423	0.015	.000492	.0046531
gdp651	−.0346201	.0092421	−3.746	0.000	−.0527342	−.0165059
inv80s	.0999582	.0223502	4.472	0.000	.0561527	.1437638
_cons	.2532421	.0792914	3.194	0.001	.0978339	.4086503
			FFP growth (tfp80)			
log_qc8	−.101905	.0330227	−3.086	0.002	−.1666284	−.0371817
qcap87	.011366	.0036251	3.135	0.002	.0042609	.0184712
human65	.0014746	.000857	1.721	0.085	−.0002051	.0031542
gdp651	−.022159	.0068844	−3.219	0.001	−.0356522	−.0086659
_cons	.1760772	.0573629	3.070	0.002	.0636479	.2885065

5.   Conclusions

Although this analysis is preliminary, the results reported in this paper suggest, quite strongly, that the positive relationship between capital account openness and productivity performance only manifests itself after the country in question has reached a certain degree of development. A plausible interpretation is that countries can only take advantage, in the net, of a greater mobility of capital once they have developed a somewhat advanced domestic financial market. I explored this interpretation by using a term that interacted the CAPOP index with standard measures of domestic financial development, including the ratio of liquid liabilities in the banking sector to GDP and the exchange rate black market premium. Broadly speaking, these results support the view that while for financially sophisticated countries an open capital account is a boon, at very low levels of local financial development, a more open capital account may have a negative effect on performance. In that sense, then, emerging markets are essentially "different" from advanced nations.

Bibliography

Alesina, A., V. Grilli, and G.-M. Milesi (1994). The Political Economy of Capital Controls. In L. Leiderman and A. Razin (eds.), *Capital Mobility: The Impact on Consumption, Investment and Growth.* Cambridge: University Press.

Barro, R., and X. Sala-i-Martin (1995). *Economic Growth.* Cambridge: MIT Press.

Bhagwati, J. (1998). The Capital Myth: The Difference Between Trade in Widgets and Trade in Dollars. *Foreign Affairs* 7(7): 7–12.

Bosworth, B., R. Dornbusch, and R. Labán (1994). *The Chilean Economy: Lessons and Challenges.* Washington, D.C.: The Brookings Institution.

Budenich, C., and G. Lefort (1997). Capital Account Regulation and Macroeconomics Policy: Two Latin American Experiences. Banco Central de Chile.

Calvo, G. (1998). The Simple Economics of Sudden Stops. *Journal of Applied Economics* 1(1): 35–54.

Campbell, J., A. Lo, and A.C. MacKinlay (1997). *The Econometrics of Financial Markets.* Princeton: Princeton University Press.

Cooper, R.N. (1998). Should Capital Account Convertibility Be a World Objective? Essays in International Finance 207. Princeton.

Cowan, K., and J. De Gregorio (1997). Exchange Rate Policies and Capital Account Management: Chile in the 1990s. Mimeo, Departamento de Ingeniería Industrial, Universidad de Chile, Santiago.

Cuddington, J. (1986). Capital Flight: Estimates, Issues and Explanations. Essays in International Finance 58. Princeton.

De Gregorio, J., S. Edwards, and R. Valdés (1998). Capital Controls in Chile: An Assessment. Presented at the Interamerican Seminar on Economics, Rio de Janeiro, Brazil. http://www.anderson.ucla.edu/faculty/sebastian.edwards/.

Dooley, M. (1995). Capital Mobility and Economic Policy. In S. Edwards (ed.), *Capital Controls, Exchange Rates, and Monetary Policy in the World Economy.* Cambridge: Cambridge University Press.

Dooley, M., D. Mathieson, and L. Rojas-Suarez (1997). Capital Mobility, and Exchange Market Intervention in Developing Countries. NBER Working Paper 6247. Cambridge, Mass.

Dornbusch, R. (1998). Capital Controls: An Idea Whose Time Is Past. Essays in International Finance 207. Princeton.

Dornbusch, R., and S. Edwards (1991). *The Macroeconomics of Populism in Latin America.* Chicago: University of Chicago Press.

Edwards, S. (1985). The Pricing of Bonds and Bank Loans in International Markets: An Empirical Analysis of Developing Countries' Foreign Borrowing. National Bureau of Economic Research Working Paper 1689, Cambridge, Mass.

Edwards, S. (1988). Financial Deregulation and Segmented Capital Markets: The Case of Korea. *World Development* 16 (January): 185–194.

Edwards, S. (1989). *Real Exchange Rates, Devaluation and Adjustment.* Cambridge, Mass: MIT Press.

Edwards, S. (1996). Why Are Latin America's Savings Rates So Low? An International Comparative Analysis. *Journal of Development Economics* 51(1): 5–44.

Edwards, S. (1998a). Capital Flows, Real Exchange Rates and Capital Controls: Some Latin American Experiences. NBER Working Paper 6000. Also in author's web page: http://www.anderson.ucla.edu/faculty/sebastian.edwards/.

Edwards, S. (1998b). Interest Rate Volatility, Contagion and Convergence: An Empirical Investigation of the Cases of Argentina, Chile and Mexico. *Journal of Applied Economics* 1(1): 55–86.

Edwards, S. (1998c). Openness, Productivity and Growth: What Do We Really Know? *The Economic Journal* 108(447): 383–398. Also in: http://www.anderson.ucla.edu/faculty/sebastian.edwards/.

Edwards, S., and A.C. Edwards (1991). *Monetarism and Liberalization: The Chilean Experiment.* Chicago: University of Chicago Press.

Edwards, S., and M.S. Khan (1985). Interest Rates in Developing Countries. *Finance and Development* 22(2): 28–31.

Edwards, S., and J. Santaella (1993). Devaluation Controversies in the Developing Countries. In M. Bordo and B. Eichengreen (eds.) *A Retrospective on the Bretton Woods System.* Chicago: University of Chicago Press.

Eichengreen, B. (1999). *Toward a New International Financial Architecture: A Practical Post-Asia Agenda.* Washington, D.C.: Institute for International Economics.

Eichengreen, B., and C. Wyplosz (1993). The Unstable EMS. *Brookings Papers on Economic Activity* (1): 51–124.

Feldstein, M., and C. Horioka (1980). Domestic Saving and International Capital Flows. *Economic Journal* 90 (June): 314–329.

Fischer, S. (1998). Capital Account Liberalization and the Role of the IMF. Essays in International Finance 207. Princeton.

Frankel, J.A. (1980). Quantifying International Capital Mobility in the 1980s. National Bureau of Economic Research Working Paper 2856, Cambridge, Mass.

Garber, P.M. (1998). Buttressing Capital Account Liberalization with Prudential Regulation and Foreign Entry. Essays in International Finance 207. Princeton.

Goldman Sachs (1997). *Emerging Markets Biweekly.* Several Issues.

Hanson, J. (1995). Opening the Capital Account: Costs, Benefits and Sequencing. In S. Edwards (ed.), *Capital Controls, Exchange Rates and Monetary Policy in the World Economy.* Cambridge: Cambridge University Press.

Haque, N.U., and P.J. Montiel (1991). How Mobile Is Capital in Developing Countries? *Finance and Development* 28(3): 38–39.

Harberger, A.C. (1980). Vignettes on the World Capital Market. *American Economic Review* 70(2): 331–337.

Ito, T., and R. Portes (1998). Dealing with the Asian Financial Crises. European Economic Perspectives, CEPR.

Kaminsky, G.L., and C.M. Reinhart (1999). The Twin Crises: The Causes of Banking and Balance of Payments Problems. *American Economic Review* 89(3): 473–500.

Klein, M., and G. Olivei (1999). Capital Account Liberalization, Financial Depth and Economic Growth. NBER Working Paper 7384. Cambridge, Mass.

Klepper, S., and E.E. Leamer (1984). Consistent Sets of Estimates for Regressions with Errors in All Variables. *Econometrica* 52(1): 163–183.

Krugman, P. (1998). Saving Asia: It's Time to Get Radical. *Fortune*, September 7: 74–80.

Krugman, P. (1999). Depression Economics Returns. *Foreign Affairs* 78(1).

Labán, R., and F. Larraín (1997). El Retorno del los Capitales privados a Chile en los Noventa: Causas, Efectos y Reacciones de Política. *Cuadernos de Economía* 34 (December): 339–362.

Massad, C. (1998a). The Liberalization of the Capital Acount: Chile in the 1990s. Essays in International Finance 207. Princeton.

Massad, C. (1998b). La Política Monetaria en Chile. *Economía Chilena* 1(1): 10–25.

McKinnon, R.I. (1973). *Money and Capital in Economic Development*. Washington, D.C.: The Brookings Institution.

McKinnon, R.I. (1991). *The Order of Economic Liberalization*. Baltimore: Johns Hopkins University Press.

McKinnon, R.I., and H. Pill (1995). Credit Liberalization and International Capital Flows. Mimeo, Department of Economics, Stanford University.

Milner, C. (1996). Discovering the Truth about Protection Rackets. *World Economy* 19(5): 517–532.

Mishkin, F. (1999). Global Financial Instability. *Journal of Economic Perspectives* 13(4): 3–20.

Montiel, P. (1994). Capital Mobility in Developing Countries: Some Measurement Issues and Empirical Estimates. *World Bank Economic Review* 8(3): 311–350.

Montiel, P., and C. Reinhart (1999). Do Capital Controls and Macroeconomics Policies Influence the Volume and Composition of Capital Flows? Evidence from the 1990s. *Journal of International Money and Finance* 18(4): 619–635.

Obstfeld, M., and K. Rogoff (1996). *Foundations of International Finance*. Cambridge, Mass: MIT Press.

Quinn, D. (1997). Correlates of Changes in International Financial Regulation. *American Political Science Review* 91(3): 531–551.

Razin, A., and A.K. Rose (1995). Business-Cycle Volatility and Openness: An Exploratory Cross-Sectional Analysis. In L. Leiderman and A. Razin (eds.), *Capital Mobility: The Impact on Consumption, Investment and Growth*. Cambridge: Cambridge University Press.

Reisen, H., and H. Yeches (1993). Time-Varying Estimates on the Openness of the Capital Account in Korea and Taiwan. *Journal of Development Economics* 41(2): 285–305.

Reisen, H., and M. Soto (2000). The Need for Foreign Savings in Post-Crisis Asia. Paper presented at the ADB-OECD Forum, July 2000.

Rodrik, D. (1998). Who Needs Capital-Account Convertibility? In S. Fischer et al., Should the IMF Pursue Capital-Account Convertibility? Essays in International Finance 207. Princeton: Princeton University.

Rogoff, K. (1999). International Institutions for Reducing Global Financial Instability. *Journal of Economic Perspectives* 13(4): 21–42.

Soto, C. (1997). Controles a los Movimientos de Capitales: Evaluación Empírica del Caso Chileno. Banco Central de Chile.

Stiglitz, J. (1999). Bleak Growth Prospects for the Developing World. *International Herald Tribune*, April 10–11: 6.

Tobin, J. (1978). A Proposal for International Monetary Reform. *Eastern Economic Journal* 4(3–4): 154–159.

Valdés-Prieto, S., and M. Soto (1996). Es el Control Selectivo de Capitales Efectivo en Chile? Su Efecto Sobre el Tipo de Cambio Real. *Cuadernos de Economía* 33 (April): 77–108.

Valdés-Prieto, S., and M. Soto (1998). The Effectiveness of Capital Controls: Theory and Evidence from Chile. *Empirica* 25(2): 133–164.

Williamson, J. (1993). Democracy and the "Washington Consensus". *World Development* 21(8): 331–336.

World Bank (1993). *Latin America a Decade After the Debt Crisis*. Washington, D.C.

Zahler, R.M. (1992). Política Monetaria en un Contexto de Apertura de la Cuenta de Capitales". Speech delivered at the 54th Meeting of Latin American Central Bank Governors, San Salvador, El Salvador, May 5.

IV.

Designing the Institutional Setup of Financial Markets

F32 G15
F34 F21 019

Philip R. Lane and Gian Maria Milesi-Ferretti

External Capital Structure: Theory and Evidence

(Selected Countries)

1. Introduction

The last three decades have witnessed large changes in the level and composition of capital flows, both between industrial economies and between industrial and developing countries. While syndicated bank lending and official flows were the most common forms of international financing to developing economies during the late 1970s and early 1980s, portfolio flows and foreign direct investment have increased substantially during the second part of the 1980s and especially the 1990s.

Understanding the determinants and implications of these shifts in the structure of international capital flows is a major challenge facing international economists. Much theoretical and empirical research has been devoted to issues related to debt flows, such as sovereign risk, optimal maturity structure of debt liabilities, rollover risk, and resolution of debt crises. However, less attention has been devoted to understanding both the driving forces and implications of equity and direct investment flows. In addition, empirical work in this area has been severely hampered by the paucity of data on equity and FDI stocks, and has mostly focused on "push" versus "pull" factors in driving capital flows and on the relative volatility of different types of flows.[1]

Remark: We are grateful to Giovanni Dell'Ariccia, Enrica Detragiache, Paolo Mauro, Miguel Savastano, Jeromin Zettelmeyer, conference participants, and to our discussant Michael Frenkel for useful suggestions. Manzoor Gill, Grace Juhn, and Charles Larkin provided excellent research assistance. Lane's work on this paper is part of a research network on "The Analysis of International Capital Markets: Understanding Europe's Role in the Global Economy," funded by the European Commission under the Research Training Network Programme (Contract No. HPRN–CT–1999–00067). The views expressed in this paper are the authors' only and do not necessarily reflect those of the International Monetary Fund.

[1]	See, for example, Calvo et al. (1993), Claessens et al. (1995), and Sarno and Taylor (1999).

In this paper, we briefly review the state of the theoretical literature on the subject, and we make use of a new dataset we have constructed (extending Lane and Milesi-Ferretti 1999) to present a series of empirical regularities concerning the composition of the stock of external liabilities in developing countries. Finally, we suggest some challenges for future theoretical and empirical research. The focus of our analysis is medium-term: we characterize broad empirical regularities, rather than attempting to identify the role played by the composition of capital flows in generating financial crises.

Some authors (see, for example, Eichengreen and Fishlow 1998) have referred to the 1990s as the "age of equity finance," as opposed to the "age of debt finance" describing the period leading up to the 1982 debt crisis.[2] Indeed, the data show a sizable increase in portfolio equity flows during the 1990s, due to both supply and demand effects. On the supply side, restrictions on cross-border equity investment have been reduced—for example, pension funds and other institutional investors in industrial countries have been granted more freedom in the allocation of their assets, and improvements in communications have reduced the cost of acquiring information on assets of foreign origin. On the demand side, financial development in both industrial countries and emerging markets has been substantial. As a result, world stock market capitalization and depth increased dramatically during the 1990s (see, for example, Tesar and Werner 1995, Tesar 1999, and Stulz 1999).

In addition to portfolio equity flows, foreign direct investment flows have also played a prominent role in external financing during the past decade. Foreign direct investment flows were also a net source of external financing during the 1970s and 1980s, in contrast to portfolio equity flows, but dramatically increased in importance during the 1990s. These developments are connected with an improvement in the overall macroeconomic policy stance in developing countries (with lower inflation and public deficits), the reduction in barriers to cross-border flows, and the wave of privatizations in both industrial and developing countries. While portfolio equity investment is strongly connected to the degree of financial development of recipient countries, direct investment flows have also been directed to less developed economies.[3] Among other issues, in the remainder of this paper we examine whether there are systematic differences between the determinants of these two types of flows.

The structure of external debt flows has also changed substantially, as already highlighted above: portfolio debt flows have played an increasingly important

---

2   See also the papers by Buch and Pierdzioch and by Hull and Tesar in this volume.

3   See the evidence in Hausmann and Fernández-Arias (2000a) and Albuquerque (2000). See also Borenzstein et al. (1998).

role, substituting for a decline in the share of syndicated bank lending. While the structure of debt flows is undoubtedly a crucial issue, in this paper we will focus mostly on the choice between debt and equity finance, rather than exploring shifts in the composition of debt finance.

The rest of this paper is organized as follows. Section 2 summarizes the theoretical literature. Section 3 briefly discusses existing empirical evidence and presents a broad set of stylized facts concerning the time-series behavior of capital flows and the cross-sectional distribution of the stock of external liabilities and its composition. Section 4 discusses the research agenda ahead and provides some conclusions.

2.   Theory

Why should the composition of capital flows and external assets and liabilities matter? Different types of capital flows have different properties with regard to features such as risk, liquidity, "lumpiness," tradability, reversibility, expropriability, and tax treatment. In addition, the composition of capital flows may influence productivity growth in the recipient country. For example, direct investment in developing countries can involve a transfer of technology and entrepreneurial skills, as well as a financial operation, while international portfolio equity flows may be useful in stimulating stock market development and improved corporate governance. A key feature that is especially important for vulnerable developing countries is that foreign direct investment and portfolio equity flows entail different risk-sharing properties between domestic and foreign residents in comparison to external debt flows. For example, if negative shocks to the domestic economy result in a real exchange rate depreciation, the burden of servicing foreign-currency-denominated external debt will be countercyclical, while returns on FDI and equity will be procyclical.

International macroeconomic theory has not fully kept pace with the evolution of international capital markets during the past decade. In the deterministic current account models of the 1970s and 1980s, the emphasis was on aggregate net flows under perfect or imperfect capital mobility. The debt crisis literature made advances in understanding the role played by sovereign risk and credit rationing in limiting debt flows but had little to say about the alternatives of foreign direct investment and portfolio equity flows.[4] Rather, the foreign direct investment lit-

---

[4]   See Eaton and Fernandez (1995), Cline (1995), and Lane (1999b) on external debt.

erature has typically abstracted from its financial dimension to focus on industrial organization and trade issues (e.g., the standard organization-location-internalization paradigm) and international portfolio flows have been analyzed as an extension to the standard optimal portfolio choice problem. Although much has been learned by treating each kind of flow in isolation, existing theory has little to say on the optimal structure of capital flows in terms of the relative balance between debt, foreign direct investment, and portfolio equity components.

A natural starting point in thinking about alternative sources of external finance is the corporate finance literature on the optimal capital structure of firms (see the surveys by Harris and Raviv [1991] and Rajan and Zingales [1995]). Under perfect information and no distortions, the Modigliani–Miller theorem proves the irrelevance of capital structure. Accordingly, this literature highlights the role of asymmetric information, agency problems, taxation, and corporate control considerations in determining the choice between equity and debt financing. The most famous illustration of the problems caused by asymmetric information is the "lemons" problem: equity will be underpriced, since investors will be suspicious of the fundamentals of any firm that is willing to sell an equity share. In addition, the choice of capital structure can be a useful signaling device in revealing information to investors. Agency problems between owners and managers can be ameliorated by appropriate financing choices: for example, a high debt load acts as a disciplining device in reducing managerial discretion. Finally, the fact that equity carries votes but debt does not means that capital structure can be used strategically to influence the outcome of corporate control contests. Overall, the literature has successfully established a small number of general principles governing capital structure decisions. However, their very generality means that a large number of potential determinants of capital structure can be identified as empirical counterparts to the theoretical propositions.

There are, however, a number of issues that limit the applicability of insights from the corporate finance literature to international capital flows. With regard to *informational asymmetries*, the corporate finance literature does not distinguish between domestic and foreign investors. To the extent that it has addressed these issues, the literature on the structure of international capital flows has emphasized asymmetric information problems that are exacerbated for foreign investors. For example, Gordon and Bovenberg (1996) argue that greenfield FDI is attractive because it is less prone to asymmetric information problems than other types of investments in which the foreign agent must rely on domestic owners for information. The Gordon–Bovenberg model has been extended by Razin et al. (1998a, 1998b, 1999) in a series of papers that study how different degrees of informational asymmetries and differences in tax treatment affect the composition of capital flows.

More generally, this literature on international capital flows focuses on two types of informational asymmetries: between foreign and domestic investors and between the controlling owner of the firm and outsiders, be they domestic or foreign. In this environment, FDI may be a way to reduce or eliminate the informational asymmetry which gives rise to the underpricing of equity, insofar as purchasing a controlling interest in a firm allows the foreign investor to eliminate informational problems. Because of the "lemons" problem referred to earlier, a high-productivity firm would prefer to issue debt rather than equity. However, under uncertainty, the existence of bankruptcy costs may lead to a preference for equity finance (a point that generally applies to the whole capital structure literature). With the development of sophisticated stock markets, the adverse selection problem would be mitigated and equity would become a more feasible means of financing.

Another important difference between domestic corporate finance and the external capital structure of countries is related to the *enforceability of claims.* Domestic financial contracts can be enforced by the legal system, whereas this may be more difficult in the case of international investments. A number of studies have emphasized differences in the level of expropriability in explaining the composition of capital flows. For instance, Cole and English (1991, 1992) argue that FDI is more subject to expropriation than debt is. However, they claim that expropriation risk is likely declining in the level of FDI (in contrast to debt), which is an argument in favor of the clustering of FDI in a few locations.[5] Albuquerque (2000) argues instead that it is FDI that is less subject to expropriation risk: it is "inalienable"—useless to domestic agents who are unable to operate the proprietary technology.

The degree to which these theoretical arguments apply depends on the sectoral allocation of foreign direct investment. Foreign direct investment in developing countries has mostly been concentrated in either capital-intensive sectors or in primary commodities. In particular, extractive industries are an important sector for FDI—they are typically very capital-intensive and large scale, and may have proprietary technology. Historically, FDI has been subject to significant expropriation risk (see Sigmund [1980] for several examples relating to Latin America). Albuquerque's theory may apply mostly to FDI in high-tech sectors, which may be sufficiently inalienable to lower expropriation risk.

---

[5]   With higher FDI, domestic consumption is higher with and without expropriation, but the long-run level of consumption under expropriation is unchanged (FDI depreciates). This implies a higher decline in consumption following expropriation and hence a lower probability of expropriation. Cole and English also argue that governments may have an incentive to opportunistically expropriate FDI during good times, whereas external debt repudiation is more likely during crises.

It is well understood that a desire to share risk provides an important motivation to use equity financing. At the level of a firm, avoiding bankruptcy costs is a powerful motivation to share risk. From the point of view of a country, the desire to smooth consumption is an important additional motivation in pursuing risk-sharing arrangements. Cole and English (1992) note that equity investment by foreign residents has more desirable properties than debt—for example, in the case of debt overhang, external debt may act as a disincentive to domestic investment, given that foreigners would capture (part of) the benefits of increases in output. More generally, equity allows for more favorable risk sharing, given that foreign investors bear part of the country risk in the event of a negative shock (the value of equity declines, while the value of debt often increases with respect to GDP in case the crisis is associated with a real depreciation).[6] However, Gertler and Rogoff (1990), Atkeson (1991) and Lane (1999a) show that moral hazard and repudiation risk limit the scale of state-contingent financing. These problems bedevil the scale of risk-sharing schemes in general. Indeed, a simple debt contract is the solution to the classic "costly state verification" problem studied by Townsend (1979).

Finally, in an interesting recent contribution Hull and Tesar (2000) present a general equilibrium model to study the implications of trade in equity and FDI. FDI improves productive efficiency by allowing countries to better exploit sectoral comparative advantage. In addition, in the absence of portfolio equity markets, FDI provides a mechanism to diversify against country risk. An important point is that it is ambiguous whether FDI and portfolio equity are complements or substitutes. On the one side, portfolio equity flows substitute for FDI by providing more efficient diversification (claims on income can be traded independently of production decisions). On the other, diversification through equity markets increases the risk-adjusted return to FDI and implies that equity and FDI flows are complements. When country-specific risk is large relative to industry-specific risk and the benefits of specialization through FDI are small, the authors find that equity trade may substitute for FDI flows.[7] The impact of portfolio insurance on production decisions has also been studied by Obstfeld (1994) and Feeney

---

6   It should be noted that state-contingent debt contracts can replicate equity arrangements quite closely, so the "debt/equity" distinction can be murky in some cases. Obstfeld and Rogoff (1996: Chapters 5–6) provide a textbook treatment of imperfections in international financial markets.

7   However, Kraay et al. (2000) provide an alternative explanation regarding the composition of capital flows between debt and equity. They emphasize that foreign equity is a worse hedge against sovereign (default) risk than foreign loans are, supporting the historical predominance of the latter in capital flows to developing countries.

(1994), amongst others, and suggests that the structure of capital flows can have significant effects on trade patterns and growth rates.[8]

a.   Empirical Determinants of the External Capital Structure

The papers cited above suggest some key factors in determining the structure of capital flows. As was mentioned above, the generality of the core theoretical principles means that plausible hypotheses can be entertained regarding a large number of potential determinants. We emphasize that it is important not to look at individual types of capital flows in isolation: we want to know whether a given determinant has similar effects on all types of inflow or has a differential effect on the external capital structure. Moreover, an important issue is whether different types of flows are complements or substitutes. For instance, FDI may bring about inflows of debt and portfolio equity, if these are complementary. Conversely, what comes in as FDI could go out as a debt or equity outflow, in particular if foreign investors hedge by borrowing in the country of destination and using the proceeds for capital repatriation.

In thinking about the various factors emphasized by the theoretical literature, the level of development (as proxied by output per capita) plays a multifaceted role. First, financial development likely means that asymmetric information problems are diminished, encouraging equity structures. Second, in less developed countries "family" firms that do not issue outside equity are more prominent. Third, we may expect the development of well-functioning financial markets to stimulate marketed liabilities (debt, portfolio equity, and FDI in the form of mergers and acquisitions) over nonmarketed liabilities (greenfield FDI): De Gregorio (1998) stresses the complementarities between financial development and financial integration.[9] However, absolute FDI inflows may still be a positive function of output per capita, since FDI is attracted by high levels of human capital and large market size. Finally, with respect to the debt/equity split, firms may grow large enough to be less exposed to bankruptcy and risk, promoting debt over equity. Since financial development is not perfectly correlated with output per capita, it is also interesting to examine the correlation between capital

---

[8]   We do not pursue the impact on production decisions in this paper, but leave it for future research. Borenzstein et al. (1998) study the impact of FDI on growth and highlight the role played by human capital in attracting FDI. Zebregs (1998) also points out that FDI may endogenously alter the production structure. Scheide (1993) finds no correlation between recourse to external capital and growth rates.

[9]   Similarly, the composition of debt should switch from bank loans to bonds.

inflows and the size of domestic financial markets (holding fixed output per capita). In the empirical work below, we include three measures of domestic financial activity: stock market capitalization, the M2/GDP ratio, and the scale of privatization revenues in relation to GDP.

How can one relate asymmetric information, which is clearly a key financial friction, to "observables"? Asymmetric information problems are likely to be more severe, the greater the "difference" between the investor and the target country. This "difference" may be related to factors such as proximity, language, "cultural" factors, and legal systems.[10] On the other hand, the problem may be mitigated by common and well-enforced accounting and legal principles, financial market sophistication and good telecommunications.[11] In the empirical work in this paper, we make a start by considering the role played by country size (in addition to output per capita and financial market development) and leave the investigation of other factors for future research. If fixed costs of acquiring (and providing) information are important, we may expect larger countries to more successfully attract information-intensive forms of finance, such as portfolio equity flows. One important issue that we do not address in this paper is whether the external capital structure of countries reflects their domestic capital structure—namely, whether factors such as the nature of the legal system and protection of minority shareholders, which are known to affect the structure of domestic capital markets, also play a role in shaping the external financing.

Trade openness is an obvious possible explanatory variable. Similar to the level of GDP per capita, trade openness plausibly has several effects on capital structure. First, trade openness may increase external vulnerability and hence the desire for risk sharing, favoring equity over debt. Second, trade openness may ameliorate asymmetric information problems, since goods trade increases familiarity and provides useful information to overseas investors. Third, trade openness plausibly reduces repudiation risk: more open economies are more vulnerable to trade sanctions and may be better able to post tradable collateral (Lane 2000a). Finally, openness in trade may also reflect a liberal policy environment that generally stimulates asset trade.[12] In the empirical work, we also examine the level

---

[10]   On the implications of differences in legal systems for the choice of financing by firms, see La Porta et al. (1998). On the role of proximity in the determination of capital flows, see Ghosh and Wolf (1998) and Portes and Rey (1999).

[11]   On the role of telecommunications in promoting capital flows, see Kim (1999) and Portes and Rey (1999). Large equity issues would make any fixed cost of information more easily paid; in addition, more information may be available.

[12]   Of course, the level of trade is endogenous to FDI in particular. In future work, we plan to address this endogeneity issue.

of natural resource exports as a determinant of capital inflows, in line with the idea that FDI may be particularly high in the natural resource sector, which is often capital-intensive.

Finally, since capital controls may have a differential impact on specific types of capital flows, we also consider a measure of foreign exchange restrictions in the empirical work below. International taxation is another factor which has been shown to affect the location of foreign direct investment, but because of the lack of comparable data for the large sample of countries we do not consider it explicitly. The composition of liabilities is also heavily influenced by the general (domestic and international) policy environment and incentive structure, as highlighted by Rogoff (1999). Deposit insurance and bailout schemes for debtholders stack the deck in favor of debt and against equity/FDI. Moreover, if a "fixed" exchange rate encourages foreign currency debt transactions, it may crowd out FDI and equity flows, although a credible peg may also encourage equity inflows by reducing exchange rate risk. We defer the investigation of these other policy measures to future work.

3.   Empirical Evidence

This section provides a broad-brush picture of empirical regularities concerning the composition of external liabilities in industrial and developing countries. Our sample includes 132 countries, of which 22 are classified as industrial (see the Appendix for a country list). We subdivide developing countries into five groupings: sub-Saharan Africa, the Middle-East and North Africa, Asia, Latin America, and transition economies. We focus on cross-country heterogeneity and on (gross) stocks rather than flows. We think that from a macroeconomic perspective this is the correct approach. The stock position is the relevant state variable in a macroeconomic model and capital flows arise to close the gap between desired and actual stock positions. From a risk-sharing perspective, the benefits of diversification are provided by holding stocks of external assets and liabilities, which generate international investment income flows and capital gains (or losses). Moreover, much of the benefits of asset trade derive from gross, rather than net, stock positions. Finally, focusing on stocks rather than flows helps reduce the "noise" inherent in the year-to-year fluctuations in flows.

Our particular emphasis is on external liabilities, rather than external assets. There are several reasons for this choice. First, the stocks of equity and FDI investment abroad are very limited for most developing countries, with a few exceptions among the high- and middle-income countries such as Singapore and Taiwan province of China. Second, estimates of debt assets held by developing countries are fraught with the problems discussed extensively in the capital flight literature.

Our regressions simply try to establish some broad empirical regularities by exploring which country features are associated with a different composition of gross external liabilities. We divide external liabilities into three categories: external debt, direct investment liabilities, and portfolio equity liabilities. For external debt, we use data from the World Bank (Global Development Finance). The methodology for the estimation of FDI and equity stocks is explained in detail in Lane and Milesi-Ferretti (1999). This methodology provides estimates of the stock of FDI calculated at book value and of the stock of equity calculated at market value. We focus on the stock position in 1997, the last year for which we have consistent data availability for the countries in our sample. We chose to use the latest available year, rather than an average over the 1990s, because of the changes in the composition of capital flows during the decade—an average would understate the current importance of equity and FDI. Our results are not significantly affected by the Asian crisis—they remain virtually unchanged when we use data for 1996.

Table A1 in the Appendix and Figures 1 and 2 provide a statistical summary of the variables used in the empirical analysis. Among the most notable features are the higher dispersion of net foreign asset positions in developing countries (the country accounting for the largest NFA position is Kuwait, the lowest the Republic of Congo) and the larger stocks of portfolio equity liabilities in industrial countries.[13] Within developing countries, Figures 1 and 2 show that countries from sub-Saharan Africa have the most negative NFA positions, largely because of their high external debt. Figures 1 and 2 also suggest differences between the cross-country distribution of FDI and equity liabilities—indeed, the correlation between the two stocks is very low (0.07 for developing countries and 0.15 for industrial countries).

The regressions (presented in Tables A2–A8) have the same structure, which is similar to the one used for FDI flows in Hausmann and Fernández-Arias (2000a). In the first table, the dependent variable is the net foreign asset position. Although our focus in this paper is on gross liabilities, it is important to understand the net position as a background for the study of the liability side of the balance sheet. We analyze total external liabilities in Table A3, the stocks of external debt, FDI liabilities, and equity liabilities in Tables A4–A6. Tables A7–A8 examine the ratio of "equity" to "debt" liabilities and the share of FDI in total

---

[13]   Indeed, the median level of portfolio equity liabilities in developing countries is zero (Table A1).

Figure 1: External Position, Regional Means, 1997

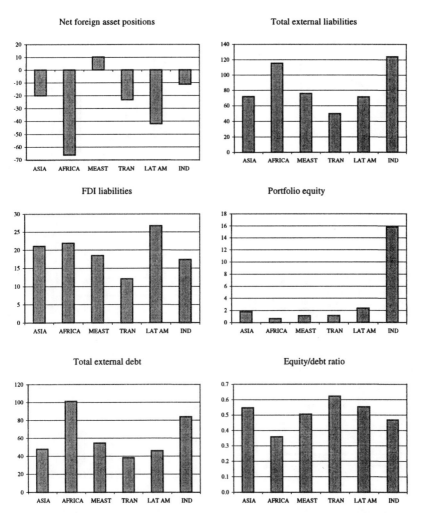

Note: All variables except equity/debt ratio are expressed as ratio to GDP (times 100).

Figure 2: External Position, Regional Medians, 1997

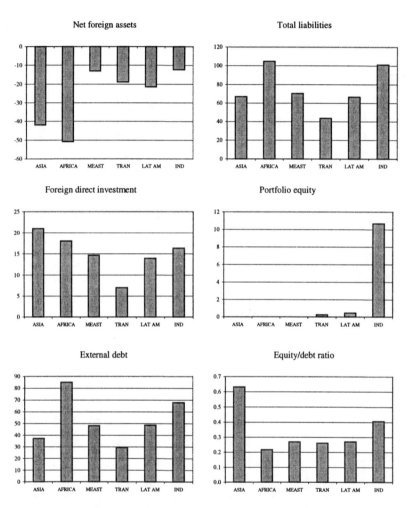

Note: All variables except equity/debt ratio are ratios of GDP (times 100).

external liabilities. We report regression results for industrial and developing countries separately; regression results for the whole sample are available from the authors.

We use as fixed controls the level of GDP per capita in dollars (in logs), country size (total GDP in dollars, in logs) and the degree of trade openness (defined as the sum of imports and exports over GDP) as well as continent dummies.[14] We then add a series of additional controls, one by one. The first two extra regressors are the share of exports of ore plus fuel as a ratio of GDP (*nat res*), and the share of privatization revenues in GDP (*privat/GDP*). The former could be a potential determinant of the stock of FDI, if FDI is attracted to resource-rich sectors. The latter captures the release of state-owned assets to private investors, stimulates domestic financial development and may attract both FDI and portfolio equity inflows. The variable *privat/GDP* is available only for developing countries and the variable *nat res* is not available for the republics of the former Soviet Union, with the exception of Estonia and Russia.

The next two regressors are the ratio of stock market capitalization to GDP (*stock mkt cap*) and the ratio of M2 to GDP (*M2/GDP*), which proxy for the degree of financial development of a country. As argued above, financial development is plausibly associated with "marketed" financial instruments, such as portfolio equity and debt liabilities, with an ambiguous impact on FDI. Although the level of financial development is correlated with GDP per capita (in our sample, the correlations of GDP per capita with M2/GDP and stock market capitalization are 0.6 and 0.4, respectively), entering these variables may provide some additional information. Finally, we also include a measure of foreign exchange controls (*FX restrictions*), described in the Appendix, to measure the impact of policy restrictions on the level and composition of external liabilities (see Eichengreen et al. [1998] for a discussion).

4. Results

The results in Table A2 indicate a positive relation between the net foreign asset position and GDP per capita, both in industrial and developing countries: richer countries are larger creditors/smaller debtors. Among industrial countries, the ratio of M2 to GDP and (in most specifications) country size are also positively correlated with the net external position: larger and more financially developed

---

14 Our regressions also feature a constant and hence exclude one dummy (the one for Latin America). This implies that the dummy coefficients represent a difference in the constant term between the group of countries in question and Latin America. The inclusion of continent dummies is a difficult choice. On the one side, the dummies control for unobserved fundamentals that differ systematically across regions. On the other, these regional differences may be highly correlated with "interesting" and observable regressors and may wipe out their individual significance. For completeness, we also present results of the basic regression excluding continent dummies.

economies have higher net foreign assets. Among developing countries, transition economies have lower external liabilities, after controlling for other NFA determinants. This is of course not surprising, given the fact that most of these economies did not become independent states until the early 1990s.[15]

Results in Table A3 clearly suggest that the ratio of gross external liabilities to GDP is higher in countries that are more open to trade, both among industrial and developing countries. In the former, external liabilities are also higher in the presence of more developed financial markets (as measured by stock market capitalization and the ratio of M2 to GDP). Taken together with the results of Table A1 and Figures 1–2, this suggests that, among industrial countries, greater openness to trade and more developed financial markets lead to more "external diversification"—that is, larger stocks of external assets and external liabilities (see also Lane 2000b). These results are indeed confirmed by similar regressions for total external assets (not reported, available from the authors). In developing countries, total external liabilities exhibit a strong negative correlation with income per capita, and a positive one with openness to trade. For this group of countries, income per capita therefore seems the crucial variable in explaining the overall net external position and the stock of external liabilities, while the results on trade openness suggests complementarities between trade in goods and trade in assets.

The results in Table A4 suggest that GDP per capita and trade openness account for an important fraction of cross-country heterogeneity in debt liabilities in developing countries: poorer and more open countries tend to have larger ratios of external debt to GDP.[16] It is important to note that a large fraction of external debt in developing countries is public or publicly guaranteed (see Table A1); in particular, sub-Saharan Africa has high debt liabilities, reflecting dependence on official financing. The same regressions using private nonguaranteed debt as the dependent variable indicate a *positive* correlation with GDP per capita, in addition to the positive relation with openness (results not reported). The fact that external debt is not correlated with domestic financial development may suggest that some degree of substitution is taking place: on the one side, a lack of domestic debt markets prompts borrowers to raise funds on international markets but on the other it hampers external financing. For industrial countries, trade openness and the degree of financial development are the most important correlates of debt liabilities; high debt liabilities are typically matched by high debt assets and mostly reflect the internationalization of the debt markets and the banking sector.

---

[15]   On determinants of capital flows to transition economies, see Garibaldi et al. (1999).

[16]   These results are in line with the findings of Lane (1999a, 2000a).

Differences between industrial and developing countries are apparent also from the regressions explaining the stock of FDI liabilities in Table A5. In industrial countries, income per capita, trade openness, and stock market capitalization explain a significant fraction of cross-country heterogeneity. In particular, the important role of the stock market variable may reflect the high proportion of FDI in industrial countries that takes the form of mergers and acquisitions, rather than greenfield investment.[17] In developing countries, trade openness and both the share of natural resource exports and the share of privatization are positively correlated with the stock of FDI liabilities.[18] It is also apparent from Table A5b that Latin American countries tend to have a larger share of FDI than other developing economies, after controlling for other determinants of FDI stocks, a finding stressed by Hausmann and Fernández-Arias (2000a).

Not surprisingly, measures of financial development are strongly correlated with the share of portfolio equity liabilities in GDP, in both industrial and developing countries (see Table A6). These findings confirm those obtained by Portes and Rey (1999) who focus on bilateral portfolio equity flows. For the developing countries group, country size is another important determinant of equity liabilities—the bigger developing countries tend to have larger equity liabilities. This result is probably explained by the fact that, on average, smaller economies are less likely to have developed stock markets. It is important to note that a significant fraction of the developing countries in our sample have no portfolio equity liabilities (52 among those included in the "basic" regression). If we exclude those countries, regression results are similar but the coefficient on trade openness becomes larger and very significant. Looking at Tables A5 and A6, the fact that privatization stimulates FDI flows to developing countries but not portfolio equity may suggest that foreign investors primarily take controlling stakes in privatized companies.

In Tables A7–A8, we turn to the composition of external liabilities. In Table A7, we examine the equity-to-debt ratio, where equity is measured as the sum of FDI and portfolio equity liabilities. The most striking finding for developing countries is the positive and significant relation between this ratio and trade openness, consistent with the notion that exposure to risk leads to a greater reli-

---

[17] Regressions with FDI assets as the dependent variable show a similar strong relation with stock market capitalization and trade openness, as well as a *positive* correlation with GDP per capita.

[18] This correlation could in part reflect imports of plant equipment for FDI purposes. Langhammer (1988) highlights the importance of this factor for "in-kind" FDI in Indonesia.

ance on equity financing.[19] For industrial countries, the equity-debt ratio has a strong positive relationship with stock market capitalization, a finding that can be explained by recalling the strong positive relation between stock market capitalization and the stocks of equity and FDI liabilities. In both developing and industrial countries, foreign exchange rate restrictions have a negative impact on the ratio, suggesting that controls have a more severe impact on equity flows than on debt flows. In developing countries, privatization revenues are also strongly associated with the equity-debt share, primarily reflecting their impact on the stock of FDI liabilities (see Table A5b). In future work, we plan to examine whether the ratio of foreign-owned equity to external debt reflects the overall domestic "gearing" of economies.

We focus on the share of FDI in total liabilities in Table A8. Some authors (e.g., Rogoff 1999) have argued that the current policy environment favors debt finance, and that much could be gained by facilitating equity diversification across countries. On the other side, Hausmann and Fernández-Arias (2000a) provocatively ask whether substantial FDI flows are really a sign of good health for a developing country. They argue that FDI flows are particularly high in relation to total private capital inflows in countries with lower GDP per capita and higher credit risk, and hence that a higher share of FDI in total capital flows is not necessarily a sign of good health. A related point is made by Albuquerque (2000), who also highlights that the share of FDI inflows in total private capital inflows is higher for countries with higher credit risk.[20] One has to take into account, however, that while FDI is the dominant fraction of *private* capital inflows to very poor countries, private capital flows are themselves a small fraction of net resource flows. Indeed, the share of private sector liabilities in total private liabilities in 1997 was below 25 percent for developing countries with GDP per capita below $2000, and close to 50 percent for those with income per capita above $2000. Hence the share of FDI in *total* capital inflows is not necessarily larger in such countries. Indeed, the unconditional correlation of GDP per capita with the FDI share in total liabilities is positive (around 0.3), while, as already pointed out by Hausmann and Fernández-Arias (2000a) and Albuquerque (2000), the unconditional correlation with the FDI share in total *private* liabilities is negative (0.41).

Tables A8 explores this point further by examining the correlates of the stock of FDI liabilities as a ratio of total external liabilities (Table A8a) and of total

---

[19]   This result is not obvious *a priori* since trade openness is positively correlated with both the stock of external debt and the stocks of equity and FDI.

[20]   Albuquerque interprets this finding as supporting his theoretical prediction that the inalienability of FDI makes it less susceptible to confiscation risk.

private liabilities (Table A8b). The dominant factor in explaining the share of FDI liabilities in total liabilities is trade openness. Table 8b instead shows that the ratio of FDI in total *private* liabilities is significantly negatively correlated with the size of the economy, while the correlation with trade openness is statistically insignificant. This ratio is very close to unity for most poor countries, which have no access to private capital markets.

We can summarize the evidence for developing countries as follows. More advanced and more open developing countries raise the most external liabilities, but this does not lead on average to larger *net* liabilities. Trade openness stimulates all forms of capital inflows but favors equity over debt flows, especially FDI. FDI is also attracted by the presence of natural resources and is positively related to the size of privatization programs. Although the share of FDI in total liabilities does not vary with the level of development, its share in private liabilities is lower in larger economies. Country size and stock market capitalization seem to be important in attracting portfolio equity. Although foreign exchange restrictions do not reduce total external liabilities, the equity-debt ratio is negatively affected by such controls. Finally, the data indicate regional differences in the structure of capital flows to developing countries. For instance, all else being equal, Latin America has a higher share of FDI liabilities in total GDP, while transition economies have lower external liabilities. Understanding the sources of these regional differences is an item on the research agenda.

5.   Policy Challenges and Conclusions

The previous section has sketched some empirical regularities concerning the cross-sectional distribution of external liabilities and their primary components. This work is very much a first step. In the future, we plan to investigate more potential determinants of the composition of external liabilities. In particular, we are interested in better understanding the roles played by the legal and regulatory environment, default and expropriation risk, and macroeconomic volatility. The political economy of the external capital structure is also relatively unexplored: what are the attitudes of domestic workers, capital owners, and politicians to different forms of capital inflows? We also plan to examine the asset side of the international balance sheet in order to obtain a more complete picture of the extent of international financial integration.

As is clear from this agenda, we think that much additional research is needed in this area. In addition to understanding the determinants of external capital structure, there has been very little work on the macroeconomic impact of different structures of external liabilities. One important issue, for example, is investi-

gating the degree of risk sharing that equity liabilities can provide. Some preliminary evidence on this subject is reported in Frankel and Rose (1996), who look at
how the composition of external liabilities influences the probability of a currency crisis, and by Hausmann and Fernández-Arias (2000b), who look in more
depth at the role of FDI in this respect, but much more work is necessary in this
area.

A second, related issue has to do with the cost of servicing external liabilities.
Insofar as equity and FDI investment require a risk premium with respect to the
rate of return on external debt, their servicing will be, ceteris paribus, more ex-
pensive even though the structure of these payouts may have more desirable cy-
clical properties. Finally, further theoretical and empirical work should provide
us with more confidence in evaluating the various policy proposals concerning
the reform of the international financial structure. There is much current discussion concerning the relative merits of different forms of capital flows. Understanding the endogenous determination of the external capital structure is a pre-
requisite for predicting the effects of policy interventions in this area.

Appendix

A.   Country List

Asia	Sub Sah. Africa	Industrial	Latin America	M. East/ N. Afr.	Transition
Bangladesh	Angola	Australia	Argentina	Algeria	Albania
Cambodia	Benin	Austria	Bolivia	Bahrain	Armenia
China	Botswana	Belgium-Lux.	Brazil	Egypt	Azerbaijan
Hong Kong S.A.R.	Burkina Faso	Canada	Chile	Iran	Belarus
India	Cameroon	Denmark	Colombia	Israel	Bulgaria
Indonesia	Cent. Afr. Rep.	Finland	Costa Rica	Jordan	Croatia
Korea	Chad	France	Dominican Rep.	Kuwait	Czech Republic
Lao People's Dem.Rep	Congo, Dem. Rep.	Germany	Ecuador	Lebanon	Estonia
Malaysia	Congo, Rep.	Greece	El Salvador	Morocco	Georgia
Myanmar	Cote d'Ivoire	Iceland	Guatemala	Oman	Hungary
Nepal	Gabon	Ireland	Haiti	Saudi Arabia	Kazakhstan
Pakistan	Gambia, The	Italy	Honduras	Syria	Kyrgyz Republic
Papua New Guinea	Ghana	Japan	Jamaica	Tunisia	Latvia
Philippines	Guinea	Netherlands	Mexico	Turkey	Lithuania
Singapore	Kenya	New Zealand	Nicaragua	Un. Ar. Em.	Macedonia, Fyr
Sri Lanka	Lesotho	Norway	Panama		Moldova
Taiwan Prov. of China	Madagascar	Portugal	Paraguay		Mongolia
Thailand	Mauritania	Spain	Peru		Poland
	Mauritius	Sweden	Trinidad & Tobago		Romania
	Namibia	Switzerland	Uruguay		Russia
	Niger	Un. Kingdom	Venezuela		Slovak Republic
	Nigeria	United States			Slovenia
	Senegal				Tajikistan
	South Africa				Turkmenistan
	Sudan				Ukraine
	Togo				Uzbekistan
	Uganda				
	Zambia				
	Zimbabwe				

B.   Data Sources

Net foreign assets: Adjusted cumulative current account balance, 1997 (see Lane and Milesi-Ferretti [1999] for methodology). Source: Lane and Milesi-Ferretti (1999) and authors' calculations.

Debt liabilities: Industrial countries: stock of portfolio debt + other liabilities (International Investment Position). Source: IMF, *Balance of Payments Statistics.* Developing countries: gross external debt. Source: World Bank, Global Development Finance.

FDI liabilities: Cumulative FDI inflows, adjusted for variations in relative prices (see Lane and Milesi-Ferretti [1999] for methodology). Source: Lane and Milesi-Ferretti (1999) and authors' calculations based on IMF, *Balance of Payments Statistics.*

Portfolio equity liabilities: cumulative flow of portfolio equity liabilities, adjusted for variations in domestic stock market indices measured in dollars (see Lane and Milesi-Ferretti [1999] for methodology). Source: Lane and Milesi-Ferretti (1999) and authors' calculations based on IMF, *Balance of Payments Statistics.*

Total external liabilities: sum of debt, FDI and portfolio equity stocks.

Log GDP per cap: log of GDP per capita in current dollars, 1997. Source: World Bank, World Development Indicators.

Log GDP: log of GDP in current dollars, 1997. Source: World Bank, World Dev. Ind.

open: imports plus exports of goods and services over GDP. Source: World Bank, World Development Indicators.

nat res: sum of exports of fuels and ore over GDP (average 1990–1997). Source: World Bank, World Development Indicators.

privat: ratio of privatization revenue to GDP (average 1994–1997). Source: World Bank, World Development Indicators.

Stockmkt cap: ratio of stock market capitalization to GDP (average 1994–1997). Source: World Bank, Global Devel. Network growth database. http://www.worldbank.org/research/growth/GDNdata.htm

M2/GDP: Ratio of M2 liabilities to GDP. Source: World Bank, World Devel. Indicators.

fx res: Index of foreign exchange restrictions (average, 1970–1996). The index is constructed by summing yearly dummy variables for the presence of (1) restrictions on capital account transactions, (2) restrictions on current account transactions, (3) multiple exchange rate practices, (4) requirement to surrender export proceeds. Source: IMF, *Exchange Arrangements and Exchange Restrictions.*

C.   Tables

Table A1: Basic Statistics (1997)

Variable	Obs.	Mean	Median	Std. dev.	Min.	Max.
			A. Industrial countries			
Net foreign assets	22	−11.2	−12.4	25.4	−61.1	37.2
External liabilities	19	123.8	101.0	66.5	42.4	273.6
External debt	19	84.0	67.8	42.2	35.1	200.0
FDI liabilities	22	17.3	16.3	11.9	0.4	46.8
Equity liabilities	22	15.8	10.7	20.9	0.0	94.7
FDI+equity/debt	19	0.5	0.4	0.3	0.1	1.0
Nat. resources exp.	22	10.1	5.5	13.1	1.4	58.6
M2	22	70.2	62.0	22.9	35.3	126.5
Stock mkt. capitaliz.	22	53.4	38.1	35.9	11.0	130.3
FX restrictions index	22	1.3	1.2	0.9	0.0	3.0
			B. Developing countries			
Net foreign assets	87	−31.2	−31.1	67.8	−363.0	368.0
External liabilities	92	77.9	70.1	45.5	6.5	241.9
External debt (total)	100	60.5	51.2	44.5	5.1	223.9
External debt (priv)	91	4.7	1.5	7.5	0	35.9
FDI liabilities	99	20.0	15.4	20.3	0.5	100.9
Equity liabilities	99	1.3	0	2.6	0	12.5
FDI+equity/debt	96	0.5	0.3	0.6	0	4.1
FDI liab./total liab.	96	24.9	20.1	17.3	0.9	80.4
FDI liab./total priv. liab.	89	77.3	87.1	25.2	15.1	100.0
Nat. resources exp.	75	26.6	12.8	29.7	0.2	99.7
Privatization	77	1.2	0.6	1.6	0.0	7.6
M2	106	35.2	26.1	26.2	4.6	165.5
Stock mkt. capitaliz.	64	32.3	14.5	52.3	0.1	252.8
FX restrictions index	108	2.6	2.9	1.1	0.0	4.0

Note: All variables are ratios of GDP (times 100), unless otherwise specified.

Table A2a: Net Foreign Assets, Industrial Countries

	(1)	(2)	(3)	(4)	(5)
log GDP per capita	34.67	35.91	35.99	33.11	41.74
	(2.38)*	(2.31)*	(2.61)*	(3.46)**	(2.93)**
log GDP	6.37	6.01	6.79	3.93	8.63
	(2.14)*	(1.89)⁺	(1.86)⁺	(1.19)	(2.01)⁺
Trade openness	0.27	0.26	0.28	0.25	0.32
	(1.42)	(1.32)	(1.36)	(1.19)	(1.41)
Nat. res.		–0.16			
		(0.35)			
Stock mkt. capitaliz.			–0.05		
			(0.28)		
M2/GDP				0.47	
				(2.83)*	
FX restrictions					6.57
					(0.92)
Constant	–454.01	–459.49	–469.93	–438.59	–564.61
	(3.03)**	(2.94)**	(3.24)**	(4.11)**	(3.41)**
Observations	22	22	22	22	22
Adjusted R^2	0.32	0.28	0.28	0.46	0.31

Table A2b: Net Foreign Assets, Developing Countries

	(1)	(2)	(3)	(4)	(5)	(6)	(7)
log GDP per capita	26.50	28.93	40.50	13.31	35.78	26.14	23.14
	(1.84)$^+$	(1.89)$^+$	(2.06)*	(2.34)*	(1.93)$^+$	(1.67)$^+$	(1.68)$^+$
log GDP	4.80	1.17	−2.58	5.51	0.83	0.93	3.87
	(0.82)	(0.17)	(0.36)	(1.75)	(0.09)	(0.12)	(0.63)
Trade openness	−0.16	−0.31	−0.49	−0.29	−0.21	−0.36	−0.40
	(0.71)	(1.15)	(1.39)	(2.26)*	(0.64)	(1.04)	(1.24)
Africa		2.00	−17.60	10.30	45.02	−0.55	3.41
		(0.07)	(0.40)	(0.65)	(1.58)	(0.02)	(0.13)
Asia		41.51	64.39	19.16	49.96	33.30	35.16
		(1.87)$^+$	(1.96)$^+$	(1.39)	(1.74)$^+$	(1.44)	(1.59)
Middle East		46.56	41.15	15.64	49.09	38.37	41.99
		(1.45)	(1.48)	(1.13)	(1.24)	(1.41)	(1.46)
Transition		38.52	36.15	38.56	29.24	38.68	46.03
		(3.42)**	(2.87)**	(4.41)**	(2.26)*	(2.95)**	(3.26)**
Nat. res.			0.23				
			(0.37)				
Privat/GDP				−0.52			
				(0.28)			
Stock mkt. capitaliz.					−0.10		
					(0.71)		
M2/GDP						0.33	
						(0.85)	
FX restrictions							−14.24
							(1.60)
Constant	−268.31	−265.32	−316.32	−190.97	−324.51	−248.61	−206.65
	(4.59)**	(4.77)**	(3.65)**	(5.35)**	(4.90)**	(4.29)**	(4.53)**
Observations	81	81	62	65	55	79	79
Adjusted R^2	0.22	0.25	0.27	0.35	0.23	0.23	0.26

Note: Estimation by OLS. *t*-statistics (calculated with heteroskedasticity-adjusted standard errors) in parentheses. $^+$ (*, **) indicates statistical significance at the 10% (5%, 1%) confidence level.

Table A3a: Total External Liabilities, Industrial Countries

	(1)	(2)	(3)	(4)	(5)
log GDP per capita	32.91	32.70	−5.93	34.60	11.06
	(0.68)	(0.68)	(0.21)	(1.09)	(0.29)
log GDP	3.94	3.89	−5.21	−2.27	−2.04
	(0.46)	(0.43)	(1.07)	(0.45)	(0.20)
Trade openness	1.50	1.50	1.46	1.36	1.33
	(4.89)**	(4.68)**	(6.12)**	(3.94)**	(6.56)**
Nat. res.		−0.07			
		(0.08)			
Stock mkt. capitaliz.			1.29		
			(3.53)**		
M2/GDP				1.47	
				(2.10)$^{+}$	
FX restrictions					−19.41
					(1.06)
Constant	−335.94	−332.62	99.55	−374.01	−8.72
	(0.77)	(0.79)	(0.36)	(1.14)	(0.03)
Observations	19	19	19	19	19
Adjusted R^2	0.20	0.14	0.67	0.44	0.18

Table A3b: Total External Liabilities, Developing Countries

	(1)	(2)	(3)	(4)	(5)	(6)	(7)
log GDP per capita	−8.73	−9.89	−13.85	−1.88	−14.61	−9.16	−5.28
	(2.03)*	(2.12)*	(2.38)*	(0.28)	(3.40)**	(1.82)⁺	(0.98)
log GDP	−6.06	−4.08	−1.66	−7.23	0.78	−4.03	−6.00
	(2.56)*	(1.26)	(0.53)	(1.74)⁺	(0.26)	(1.19)	(1.72)⁺
Trade openness	0.36	0.49	0.49	0.47	0.54	0.47	0.51
	(3.13)**	(4.01)**	(4.12)**	(2.79)**	(4.77)**	(3.56)**	(4.08)**
Africa		19.48	15.72	21.28	−19.23	20.96	20.91
		(1.41)	(0.99)	(0.97)	(1.60)	(1.48)	(1.47)
Asia		−14.13	−21.87	−6.86	−40.89	−12.10	−7.66
		(1.05)	(2.08)*	(0.44)	(4.39)**	(0.79)	(0.55)
Middle East		4.27	−4.32	−1.19	−8.88	2.98	5.44
		(0.42)	(0.39)	(0.11)	(0.81)	(0.24)	(0.56)
Transition		−39.44	−36.38	−44.95	−33.96	−38.43	−42.57
		(4.04)**	(3.45)**	(4.37)**	(3.10)**	(3.84)**	(4.33)**
Nat. res.			0.22				
			(1.18)				
Privat/GDP				1.33			
				(0.65)			
Stock mkt. capitaliz.					0.06		
					(0.64)		
M2/GDP						0.04	
						(0.18)	
FX restrictions							7.21
							(1.50)
Constant	177.62	166.07	170.07	141.94	155.67	159.55	130.58
	(6.56)**	(5.46)**	(5.56)**	(3.35)**	(5.17)**	(5.16)**	(3.74)**
Observations	87	87	64	66	53	84	85
Adjusted R^2	0.22	0.43	0.41	0.41	0.43	0.40	0.44

Note: Estimation by OLS. *t*-statistics (calculated with heteroskedasticity-adjusted standard errors) in parentheses. ⁺ (*, **) indicates statistical significance at the 10% (5%, 1%) confidence level.

Table A4a: External Debt, Industrial Countries

	(1)	(2)	(3)	(4)	(5)
log GDP per capita	18.22	17.11	0.04	19.10	18.26
	(0.80)	(0.77)	(0.00)	(1.15)	(0.94)
log GDP	2.85	2.61	−1.43	−0.38	2.86
	(0.63)	(0.53)	(0.47)	(0.15)	(0.46)
Trade openness	0.94	0.94	0.92	0.87	0.94
	(6.67)**	(6.68)**	(3.94)**	(4.38)**	(7.42)**
Nat. res.		−0.39			
		(0.77)			
Stock mkt. capitaliz.			0.60		
			(1.66)		
M2/GDP				0.77	
				(1.81)[+]	
FX restrictions					0.03
					(0.00)
Constant	−185.15	−167.57	18.73	−204.98	−185.72
	(0.89)	(0.85)	(0.08)	(1.21)	(1.05)
Observations	19	19	19	19	19
Adjusted R^2	0.18	0.13	0.41	0.33	0.12

Table A4b: External Debt, Developing Countries

	(1)	(2)	(3)	(4)	(5)	(6)	(7)
log GDP per capita	−13.89	−14.00	−14.79	−9.40	−10.37	−13.44	−10.81
	(3.48)**	(3.27)**	(3.15)**	(1.76)+	(3.19)**	(2.81)**	(2.07)*
log GDP	−5.62	−2.32	−2.00	−4.39	−2.52	−1.72	−4.28
	(2.46)*	(0.80)	(0.82)	(1.21)	(1.11)	(0.59)	(1.32)
Trade openness	0.17	0.27	0.22	0.31	0.22	0.30	0.30
	(2.03)*	(2.69)**	(2.27)*	(2.24)*	(2.61)*	(2.71)**	(2.92)**
Africa		32.92	29.33	21.73	10.81	33.59	32.18
		(2.75)**	(2.08)*	(1.28)	(0.69)	(2.81)**	(2.64)**
Asia		−11.08	−13.22	−7.29	−14.13	−8.74	−6.84
		(1.02)	(1.47)	(0.57)	(1.80)+	(0.70)	(0.59)
Middle East		12.20	6.94	4.48	6.04	16.30	15.15
		(1.41)	(0.66)	(0.45)	(0.61)	(1.61)	(1.80)+
Transition		−22.17	−15.97	−24.58	−12.00	−23.13	−23.45
		(2.95)**	(2.00)*	(2.98)**	(1.40)	(2.89)**	(3.06)**
Nat. res.			0.19				
			(1.22)				
Privat/GDP				−0.96			
				(0.53)			
Stock mkt. capitaliz.					−0.05		
					(0.57)		
M2/GDP						−0.14	
						(0.83)	
FX restrictions							7.20
							(1.92)+
Constant	204.87	164.61	167.96	153.21	145.79	156.94	140.07
	(7.21)**	(6.14)**	(6.19)**	(4.13)**	(5.19)**	(5.65)**	(4.64)**
Observations	95	95	68	70	57	92	93
Adjusted R^2	0.25	0.43	0.44	0.34	0.30	0.41	0.44

Note: Estimation by OLS. *t*-statistics (calculated with heteroskedasticity-adjusted standard errors) in parentheses. + (*, **) indicates statistical significance at the 10% (5%, 1%) confidence level.

Table A5a: Foreign Direct Investment Liabilities, Industrial Countries

	(1)	(2)	(3)	(4)	(5)
log GDP per capita	−10.64	−11.68	−16.12	−11.06	−16.90
	(2.33)*	(2.46)*	(3.58)**	(2.00)$^+$	(2.40)*
log GDP	0.32	0.62	−1.39	−0.33	−1.68
	(0.18)	(0.33)	(0.85)	(0.18)	(0.65)
Trade openness	0.23	0.24	0.22	0.22	0.18
	(2.43)*	(2.53)*	(2.39)*	(2.45)*	(1.63)
Nat. res.		0.13			
		(1.02)			
Stock mkt. capitaliz.			0.21		
			(4.07)**		
M2/GDP				0.13	
				(1.31)	
FX restrictions					−5.82
					(1.42)
Constant	107.63	112.18	173.55	111.76	205.45
	(1.85)$^+$	(1.85)$^+$	(2.97)**	(1.65)	(1.91)$^+$
Observations	22	22	22	22	22
Adjusted R^2	0.23	0.21	0.55	0.25	0.30

Table A5b: Foreign Direct Investment Liabilities, Developing Countries

	(1)	(2)	(3)	(4)	(5)	(6)	(7)
log GDP per capita	1.32	−0.19	0.36	3.29	−3.10	0.46	0.74
	(0.49)	(0.06)	(0.10)	(0.79)	(0.87)	(0.15)	(0.23)
log GDP	−1.50	−1.47	−1.62	−2.27	1.34	−1.76	−1.30
	(1.00)	(0.84)	(0.88)	(1.18)	(0.56)	(0.95)	(0.75)
Trade openness	0.21	0.26	0.27	0.22	0.31	0.25	0.26
	(4.21)**	(3.96)**	(3.70)**	(2.61)*	(3.07)**	(3.42)**	(3.88)**
Asia		−10.32	−14.47	−6.52	−19.05	−9.60	−8.49
		(1.62)	(2.11)*	(0.80)	(2.43)*	(1.49)	(1.34)
Africa		−10.26	−7.86	−5.34	−23.31	−8.73	−8.62
		(1.53)	(1.08)	(0.70)	(3.26)**	(1.24)	(1.23)
Middle East		−9.84	−17.14	−6.44	−15.50	−10.75	−10.36
		(1.51)	(2.49)*	(0.88)	(2.72)**	(1.51)	(1.62)
Transition		−20.40	−22.95	−22.53	−25.31	−19.45	−22.00
		(3.38)**	(3.64)**	(3.48)**	(3.38)**	(3.13)**	(3.83)**
Nat. res.			0.21				
			(2.22)*				
Privat/GDP				2.56			
				(2.08)*			
Stock mkt. capitaliz.					0.02		
					(0.54)		
M2/GDP						0.03	
						(0.33)	
FX restrictions							−0.41
							(0.22)
Constant	11.44	29.75	22.60	12.02	23.24	27.42	21.97
	(1.18)	(2.08)*	(1.35)	(0.51)	(1.57)	(1.85)	(1.42)
Observations	92	92	67	71	57	89	90
Adjusted R^2	0.21	0.30	0.45	0.30	0.41	0.28	0.35

Note: Estimation by OLS. *t*-statistics (calculated with heteroskedasticity-adjusted standard errors) in parentheses. + (*, **) indicates statistical significance at the 10% (5%, 1%) confidence level.

Table A6a: Portfolio Equity Liabilities, Industrial Countries

	(1)	(2)	(3)	(4)	(5)
log GDP per capita	17.59	19.12	6.37	15.77	2.66
	(0.94)	(0.93)	(0.59)	(1.36)	(0.21)
log GDP	0.99	0.55	−2.52	−1.88	−3.78
	(0.38)	(0.17)	(0.86)	(0.68)	(0.87)
Trade openness	0.08	0.06	0.07	0.04	−0.02
	(0.47)	(0.34)	(0.65)	(0.27)	(0.19)
Nat. res.		−0.20			
		(0.62)			
Stock mkt. capitaliz.			0.42		
			(2.49)*		
M2/GDP				0.55	
				(1.83)[+]	
FX restrictions					−13.88
					(1.64)
Constant	−177.32	−184.03	−42.32	−159.26	56.09
	(1.07)	(1.06)	(0.41)	(1.40)	(0.43)
Observations	22	22	22	22	22
Adjusted R^2	−0.05	−0.10	0.39	0.28	0.11

Table A6b: Portfolio Equity Liabilities, Developing Countries

	(1)	(2)	(3)	(4)	(5)	(6)	(7)
log GDP per capita	0.38	0.46	0.30	0.75	−0.08	0.48	0.55
	(1.48)	(1.41)	(0.66)	(1.81)⁺	(0.18)	(1.37)	(1.38)
log GDP	0.68	0.81	1.16	0.80	1.05	0.81	0.78
	(3.41)**	(3.48)**	(3.66)**	(2.93)**	(2.90)**	(3.38)**	(2.87)**
Trade openness	0.02	0.02	0.03	0.01	0.02	0.02	0.02
	(1.72)⁺	(1.96)⁺	(2.43)*	(0.94)	(1.49)	(1.86)⁺	(2.06)*
Africa		0.40	0.87	0.68	−1.71	0.47	0.47
		(0.51)	(0.90)	(0.67)	(1.48)	(0.59)	(0.59)
Asia		−1.11	−1.97	−0.80	−3.07	−1.02	−0.93
		(1.09)	(1.48)	(0.74)	(2.24)*	(0.98)	(0.84)
Middle East		−1.63	−1.23	−1.36	−1.77	−1.72	−1.62
		(1.91)⁺	(1.25)	(1.45)	(1.74)⁺	(1.90)⁺	(1.85)⁺
Transition		−0.65	−0.42	−0.62	−0.34	−0.60	−0.69
		(0.83)	(0.43)	(0.70)	(0.36)	(0.73)	(0.89)
Nat. res.			−0.01				
			(1.18)				
Privat/GDP				0.14			
				(1.17)			
Stock mkt. capitaliz.					0.04		
					(4.26)**		
M2/GDP						0.00	
						(0.22)	
FX restrictions							0.09
							(0.32)
Constant	−9.04	−10.45	−13.16	−12.14	−9.25	−10.75	−11.16
	(5.24)**	(4.97)**	(4.52)**	(4.31)**	(3.04)**	(4.76)**	(4.68)**
Observations	92	92	65	68	53	89	90
Adjusted R^2	0.37	0.40	0.41	0.34	0.50	0.39	0.39

Note: Estimation by OLS. *t*-statistics (calculated with heteroskedasticity-adjusted standard errors) in parentheses.
⁺ (*, **) indicates statistical significance at the 10% (5%, 1%) confidence level.

Table A7a: Equity to Debt Liabilities Ratio, Industrial Countries

	(1)	(2)	(3)	(4)	(5)
log GDP per capita	−0.08	−0.05	−0.25	−0.08	−0.35
	(0.41)	(0.24)	(1.57)	(0.41)	(1.74)
log GDP	0.01	0.01	−0.03	0.00	−0.06
	(0.13)	(0.24)	(0.85)	(0.06)	(1.46)
Trade openness	0.00	0.00	0.00	0.00	0.00
	(0.42)	(0.48)	(0.58)	(0.33)	(0.43)
Nat. res.		0.01			
		(2.40)*			
Stock mkt. capitaliz.			0.01		
			(2.42)*		
M2/GDP				0.00	
				(0.80)	
FX restrictions					−0.23
					(2.34)*
Constant	1.15	0.63	3.04	1.09	5.08
	(0.55)	(0.30)	(1.85)[+]	(0.51)	(2.14)[+]
Observations	19	19	19	19	19
Adjusted R^2	−0.18	−0.13	0.33	−0.21	0.11

Table A7b: Equity to Debt Liabilities Ratio, Developing Countries

	(1)	(2)	(3)	(4)	(5)	(6)	(7)
log GDP per capita	0.07	0.05	0.03	0.18	−0.04	0.05	0.01
	(1.18)	(0.67)	(0.35)	(1.89)[+]	(0.50)	(0.65)	(0.18)
log GDP	0.05	0.05	0.06	0.02	0.14	0.05	0.07
	(1.14)	(1.22)	(1.17)	(0.47)	(2.58)*	(1.14)	(1.65)
Trade openness	0.01	0.01	0.01	0.00	0.01	0.01	0.01
	(3.73)**	(3.38)**	(3.10)**	(1.97)[+]	(2.59)*	(3.09)**	(3.21)**
Africa		−0.15	−0.24	0.03	−0.27	−0.15	−0.14
		(1.20)	(1.59)	(0.21)	(1.92)	(1.13)	(1.10)
Asia		−0.17	−0.20	0.04	−0.43	−0.16	−0.20
		(1.11)	(0.97)	(0.22)	(2.55)*	(0.94)	(1.17)
Middle East		−0.16	−0.22	−0.07	−0.22	−0.16	−0.18
		(1.06)	(1.80)[+]	(0.46)	(1.59)	(0.91)	(1.28)
Transition		−0.25	−0.24	−0.24	−0.39	−0.24	−0.21
		(2.09)*	(1.47)	(2.04)*	(2.69)*	(1.83)[+]	(1.65)
Nat. res.			0.00				
			(1.06)				
Privat/GDP				0.06			
				(2.99)**			
Stock mkt. capitaliz.					0.00		
					(0.60)		
M2/GDP						0.00	
						(0.04)	
FX restrictions							−0.08
							(1.72)[+]
Constant	−0.87	−0.64	−0.68	−1.32	−0.94	−0.64	−0.34
	(3.84)**	(2.00)*	(1.46)	(2.97)**	(2.03)*	(1.81)[+]	(0.91)
Observations	86	86	64	65	53	83	85
Adjusted R^2	0.37	0.38	0.36	0.33	0.45	0.35	0.39

Note: Estimation by OLS. *t*-statistics (calculated with heteroskedasticity-adjusted standard errors) in parentheses. [+] (*, **) indicates statistical significance at the 10% (5%, 1%) confidence level.

Table A8a: FDI Liabilities to Total Liabilities, Developing Countries

	(1)	(2)	(3)	(4)	(5)	(6)	(7)
log GDP per capita	2.44	0.64	1.06	4.05	−2.25	1.05	−0.04
	(1.01)	(0.23)	(0.32)	(1.07)	(0.72)	(0.34)	(0.01)
log GDP	0.36	0.54	0.29	−0.59	3.76	0.24	1.45
	(0.24)	(0.32)	(0.15)	(0.28)	(1.64)	(0.14)	(0.85)
Trade openness	0.12	0.15	0.17	0.11	0.21	0.15	0.15
	(3.09)**	(2.75)**	(2.50)*	(1.49)	(2.47)*	(2.38)*	(2.71)**
Africa		−11.78	−16.27	−7.45	−14.74	−11.73	−10.35
		(2.52)*	(3.37)**	(1.15)	(2.31)*	(2.42)*	(2.17)*
Asia		−9.14	−7.76	−3.56	−19.34	−8.31	−10.26
		(1.62)	(1.08)	(0.55)	(2.93)**	(1.35)	(1.61)
Middle East		−8.38	−12.29	−5.58	−9.14	−8.29	−9.33
		(1.40)	(2.42)*	(0.95)	(1.44)	(1.20)	(1.68)
Transition		−9.74	−13.12	−9.17	−17.08	−8.40	−11.40
		(1.84)+	(2.25)*	(1.60)	(2.94)**	(1.48)	(2.37)*
Nat. res.			0.10				
			(1.53)				
Privat/GDP				1.65			
				(1.30)			
Stock mkt. capitaliz.					0.01		
					(0.21)		
M2/GDP						0.00	
						(0.03)	
FX restrictions							−2.72
							(1.65)
Constant	−4.45	12.78	9.84	−0.83	0.02	13.27	15.81
	(0.43)	(0.95)	(0.58)	(0.03)	(0.00)	(0.90)	(1.21)
Observations	87	87	64	66	53	84	85
Adjusted R^2	0.13	0.15	0.27	0.11	0.25	0.13	0.26

Table A8b: FDI Liabilities to Total Private Liabilities, Developing Countries

	(1)	(2)	(3)	(4)	(5)	(6)	(7)
log GDP per capita	−2.08	−6.83	−5.53	−7.06	−10.14	−5.71	−6.85
	(0.55)	(1.41)	(1.14)	(1.47)	(1.73)⁺	(1.15)	(1.34)
log GDP	−7.33	−7.13	−7.48	−7.27	−6.21	−7.85	−7.02
	(3.09)**	(2.79)**	(2.58)*	(2.61)*	(1.72)⁺	(3.18)**	(2.60)*
Trade openness	−0.08	−0.02	−0.03	−0.03	−0.04	−0.05	−0.02
	(1.20)	(0.32)	(0.29)	(0.42)	(0.35)	(0.70)	(0.31)
Africa		−19.68	−29.72	−10.37	−27.46	−19.64	−19.40
		(2.31)*	(3.04)**	(1.34)	(1.90)⁺	(2.34)*	(2.27)*
Asia		−12.13	−10.42	−10.35	−15.77	−9.72	−12.21
		(1.57)	(1.18)	(1.27)	(1.46)	(1.31)	(1.54)
Middle East		5.85	−1.03	5.01	1.83	7.35	5.96
		(0.72)	(0.12)	(0.53)	(0.16)	(0.67)	(0.72)
Transition		−11.54	−15.05	−10.29	−11.75	−7.65	−11.79
		(2.05)*	(1.89)⁺	(1.50)	(1.72)⁺	(1.61)	(2.06)*
Nat. res.			0.28				
			(2.44)*				
Privat/GDP				−0.24			
				(0.22)			
Stock mkt. capitaliz.					0.00		
					(0.02)		
M2/GDP						−0.05	
						(0.26)	
FX restrictions							−0.37
							(0.16)
Constant	166.14	204.26	192.65	207.60	221.19	205.74	204.18
	(11.13)**	(9.77)**	(7.05)**	(9.14)**	(7.04)**	(9.94)**	(8.51)**
Observations	80	80	58	64	47	78	79
Adjusted R^2	0.31	0.36	0.36	0.40	0.35	0.40	0.35

Note: Estimation by OLS. *t*-statistics (calculated with heteroskedasticity-adjusted standard errors) in parentheses. ⁺ (*, **) indicates statistical significance at the 10% (5%, 1%) confidence level.

Bibliography

Albuquerque, R. (2000). The Composition of International Capital Flows: Risk-Sharing through Foreign Direct Investment. Mimeo, University of Rochester.

Atkeson, A. (1991). International Lending with Moral Hazard and the Risk of Repudiation. *Econometrica* 59 (July): 1069–1089.

Borensztein, E., J. De Gregorio, and J.-W. Lee (1998). How Does Foreign Direct Investment Affect Economic Growth? *Journal of International Economics* 45 (June): 115–135.

Buch, C.M., and C. Pierdzioch (2001). The Growth and Volatility of International Capital Flows: Reconciling the Evidence. This volume.

Calvo, G., L. Leiderman, and C. Reinhart (1993). Capital Inflows and Real Exchange Rate Appreciation in Latin America: The Role of External Factors. *International Monetary Fund Staff Papers* 40 (March): 108–151.

Claessens, S., M. Dooley, and A. Warner (1995). Portfolio Capital Flows: Hot or Cold? *World Bank Economic Review* 9 (January): 153–174.

Cline, W.R. (1995). *International Debt Reexamined.* Washington, D.C.: Institute for International Economics.

Cole, H.L., and W.B. English (1991). Expropriation and Direct Investment. *Journal of International Economics* 30 (May): 201–227.

Cole, H.L., and W.B. English (1992). Two-Sided Expropriation and Equity Contracts. *Journal of International Economics* 33 (May): 77–104.

De Gregorio, J. (1998). Financial Integration, Financial Development and Growth. Mimeo, Universidad de Chile, July.

Eaton, J., and R. Fernández (1995). Sovereign Debt. In O. Grossman and K. Rogoff (eds.), *Handbook of International Economics*. Vol. 3. Amsterdam: Elsevier Science Publishers.

Eichengreen, B., and A. Fishlow (1998). Contending with Capital Flows: What Is Different About the 1990s? In M. Kahler (ed.), *Capital Flows and Financial Crises*. Ithaca: Cornell University Press for the Council on Foreign Relations.

Eichengreen, B., M. Mussa, G. Dell'Ariccia, E. Detragiache, G.M. Milesi-Ferretti, and A. Tweedie (1998). Capital Account Liberalization: Theoretical and Practical Aspects. IMF Occasional Paper 182, Washington, D.C.

Feeney, J.-A. (1994). Goods and Asset Market Interdependence in a Risky World. *International Economic Review* 35 (August): 551–563.

Frankel, J., and A.K. Rose (1996). Currency Crashes in Emerging Markets: An Empirical Treatment. *Journal of International Economics* 41 (November): 351–366.

Gastanaga, V.M., J. Nugent, and B. Pashamova (1998). Host Country Reforms and FDI Inflows: How Much Difference Do They Make? *World Development* 26 (July): 1299–1314.

Garibaldi, P., N. Mora, R. Sahay, and J. Zettelmeyer (1999). What Moves Capital to Transition Economies? Mimeo, International Monetary Fund, Washington, D.C.

Gertler, M., and K. Rogoff (1990). North-South Lending and Endogenous Domestic Capital Market Inefficiencies. *Journal of Monetary Economics* 26 (October): 245–266.

Ghosh, S., and H. Wolf (1998). The Geography of Capital Flows. Mimeo, New York University.

Gordon, R.H., and A.L. Bovenberg (1996). Why Is Capital So Immobile Internationally? Possible Explanations and Implications for Capital Income Taxation. *American Economic Review* 86 (December): 1057–1075.

Harris, M., and A. Raviv (1991). The Theory of the Capital Structure. *Journal of Finance* 46(1): 297–355.

Hausmann, R., and E. Fernández-Arias (2000a). Foreign Direct Investment: Good Cholesterol? Mimeo, Inter-American Development Bank, Washington, D.C.

Hausmann, R., and E. Fernández-Arias (2000b). Is FDI a Safer Form of Financing? Mimeo, Inter-American Development Bank, Washington, D.C.

Hull, L., and L. Tesar (2000). Risk, Specialization and the Composition of International Capital Flows. Mimeo, University of Michigan, Ann Arbor.

Kim, H.-M. (1999). *Globalization of International Financial Markets: Causes and Consequences.* Brookfield, USA: Ashgate.

Kraay, A., N. Loayza, L. Serven, and J. Ventura (2000). Country Portfolios. NBER Working Paper 7795, Cambridge, Mass.

La Porta, R., F. Lopez-de-Silanes, A. Shleifer, and R.W. Vishny (1998). Law and Finance. *Journal of Political Economy* 106 (December): 1113–1155.

Lane, P.R. (1999a). North-South Lending under Moral Hazard and Repudiation Risk. *Review of International Economics* 7 (February): 50–58.

Lane, P.R. (1999b). Empirical Perspectives on Long-Term External Debt. Mimeo, Trinity College, Dublin.

Lane, P.R. (2000a). International Trade and Economic Convergence: The Credit Channel. *Oxford Economic Papers*, forthcoming.

Lane, P.R. (2000b). International Investment Positions: A Cross-Sectional Analysis. *Journal of International Money and Finance* 19(4): 513–534.

Lane, P.R., and G.M. Milesi-Ferretti (1999). The External Wealth of Nations. CEPR Discussion Paper 2031, London. Forthcoming in *Journal of International Economics*.

Langhammer, R. (1988). Financing of Foreign Direct Investment and Trade Flows: The Case of Indonesia. *Bulletin of Indonesian Economic Studies* 24 (April): 97–114.

Lipsey, R.E. (1999). The Role of Foreign Direct Investment in International Capital Flows. NBER Working Paper 7094, Cambridge, Mass.

Obstfeld, M. (1994). Risk Diversification and Growth. *American Economic Review* 84 (December): 1310–1329.

Obstfeld, M., and K. Rogoff (1996). *Foundations of International Macroeconomics.* Cambridge, Mass.: MIT Press.

Portes, R., and H. Rey (1999). The Determinants of Cross-Border Equity Flows. CEPR Discussion Paper 2225, London. Forthcoming in *European Economic Review*.

Rajan, R., and L. Zingales (1995). What Do We Know About Capital Structure? Some Evidence from International Data. *Journal of Finance* 50(5): 1421–1460.

Razin, A., E. Sadka, and C.-W. Yuen (1998a). A Pecking-Order of Capital Flows and International Tax Principles. *Journal of International Economics* 44 (February): 45–68.

Razin, A., E. Sadka, and C.-W. Yuen (1998b). Capital Flows with Debt- and Equity-Financed Investment: Equilibrium Structure and Efficiency Implications. IMF Working Paper 98/159, Washington, D.C.

Razin, A., E. Sadka, and C.-W. Yuen (1999). An Analytical Reexamination of the Home Bias Hypothesis. Mimeo, Tel Aviv University.

Rogoff, K. (1999). International Institutions for Reducing Global Financial Instability. *Journal of Economic Perspectives* 13(4): 21–42.

Sarno, L., and M. Taylor (1999). Hot Money, Accounting Labels and the Permanence of Capital Flows to Developing Countries: An Empirical Investigation. *Journal of Development Economics* 59 (August): 337–364.

Scheide, J. (1993). Does Economic Growth Depend on External Capital? Some Evidence from 116 Countries. *Applied Economics* 25 (March): 369–377.

Sigmund, P. (1980). *Multinationals in Latin America.* Madison: The University of Wisconsin Press.

Stulz, R. (1999). International Portfolio Flows and Security Markets. In: M. Feldstein (ed.), *International Capital Flows.* Chicago: University of Chicago Press for NBER.

Tesar, L. (1999). The Role of Equity in International Capital Flows. In: M. Feldstein (ed.), *International Capital Flows.* Chicago: University of Chicago Press for NBER.

Tesar, L., and I. Werner (1995). Home Bias and High Turnover. *Journal of International Money and Finance* 14(4): 467–492.

Townsend, R.M. (1979). Optimal Contracts and Competitive Markets With Costly State Verification. *Journal of Economic Theory* 21 (October): 265–293.

Wheeler, D., and A. Mody (1992). International Investment Location Decisions: The Case of US Firms. *Journal of International Economics* 33 (August): 57–76.

Zebregs, H. (1998). Can the Neoclassical Model Explain the Distribution of Foreign Direct Investment Across Developing Countries? IMF Working Paper 98/139, Washington, D.C.

Comment on
Philip R. Lane and Gian Maria Milesi-Ferretti

Michael Frenkel

1. The Data Set

It is a great achievement of Philip Lane and Gian Maria Milesi-Ferretti to have put together a comprehensive data set on the stock of foreign debt, as well as on the stock of foreign assets and liabilities related to foreign portfolio and foreign direct investment (FDI), for so many industrial and developing countries. Since the data set is available from the website of the IMF, it will certainly soon represent the basis of a number of studies on the characteristics, the determinants, the structure, and the effects of international capital flows.

Lane and Milesi-Ferretti emphasize at the beginning of their paper that big changes have occurred in the level and the composition of capital flows over the last three decades. Referring to other publications, they point out that, in particular, there was a shift in the structure of capital flows from both bank lending and official lending in the 1970s and early 1980s to equity, i.e., portfolio investment and FDI, in the second half of the 1980s and in the 1990s. If this is indeed the case, it is important to understand both the determinants and the implications of this shift. In their paper, Lane and Milesi-Ferretti try to present a first study on these issues. They also aim to examine the empirical evidence on the driving forces of different forms of external financing and whether equity investment and debt financing have been mostly complements or substitutes. If the hypothesis according to which FDI is followed by inflows of debt and portfolio investment can be supported by empirical evidence and if we understand the conditions under which this effect takes place, this would lead to important recommendations for policymakers.

Lane and Milesi-Ferretti emphasize that their work is very preliminary. This is not a disadvantage because at the beginning of such work the main focus is on the creation of the data set itself and, at most, on taking a very first look at econometric results.

2.   From Theory to Empirical Evidence

Lane and Milesi-Ferretti begin their paper by reviewing the theoretical literature on some issues of international capital flows. As regards the literature on corporate finance, they stress the importance of asymmetric information, agency problems, and corporate control in determinng the choice between equity and debt financing. As regards the literature on the structure of international capital flows, they emphasize two types of asymmetric information that cause particular problems for foreign investors and stress the role of risk-sharing incentives for FDI. These are important determinants only too often left aside for some ultimately trade-related variables.

   One of the main problems in basing the empirical research on international capital flows on the type of theoretical considerations which Lane and Milesi-Ferretti discuss is that it is difficult to find observable economic variables for these considerations. Therefore, in order to test functions of international capital flows, one is often left with traditional variables basically related to GDP and trade. Unfortunately, Lane and Milesi-Ferretti have apparently had to go the same way. They argue that the traditional variables are proxies for the determinants they discuss. Most of the variables they use as determinants of the different types of capital flows are relatively familiar (e.g., GDP, per capita GDP, and openness). The only exception is stock market capitalization, which appears as an explanatory variable in some of their regressions.

3.   A Closer Look at the Empirical Results

The empirical analysis in the paper does not use the complete data set that Lane and Milesi-Ferretti put together for foreign assets and liabilities. Instead, it only uses data points for 1997 and, thus, is exclusively a cross-section analysis of outstanding debt and the stock of equity. The year 1997 is used because this is the last year for which a complete set of data is available for the group of countries used in the analysis.

   The use of the cross-section analysis and the year 1997 is not without problems. First, such an analysis does not permit investigating the question whether a shift occurred in the type of financing in the 1980s. Second, it cannot be used to examine whether equity flows induce debt flows or what comes in when FDI goes out. Third, since time-series analyses have to be used to analyze whether equity investment and debt financing have been mostly complements or substitutes, this is another question that will have to be the subject of future research. Fourth, although it is sensible to use capital stock data instead of capital flow data in the empirical analysis because flows are extremely volatile, the data for 1997 and thus the econometric analysis may be affected by valuation effects resulting from the Asian crisis in this year. Fifth, another problem using the year of the Asian crisis could be that the stock adjustment was affected in some countries by temporary changes in investors' attitude towards foreign capital flows. Therefore, both valuation and adjustment effects could have resulted from this crisis.

Although the paper presents a number of estimates (close to one hundred regressions), a fairly high number of estimates do not turn out to be significant. This problem can be seen in several of the tables, but it is already demonstrated by the first set of results in Tables A2a and A2b, which focus on the explanation of the level of external assets; of all the variables used in these first regressions, basically only the estimates of the per capita GDP coefficient are significant and this is not even the case for all country groups (see, for example, Table A2b). The estimates for external liabilities as shown in Table A3a reveal that the only significant determinant for industrial countries seems to be the openness of the economy.

In my view, several regressions may involve multicollinearity problems. For example, stock market capitalization can be expected to be high in countries with high per capita GDP. In addition, many empirical studies show that size and openness are negatively correlated. Therefore, one needs to be careful in using both variables as explanatory variables in the same equation.

I could also imagine that some of the results reported by Lane and Milesi-Ferretti could be interpreted differently. For example, Table A5b shows that privatization is a significant determinant of FDI flows to developing countries, while Table A6b indicates that it is not a significant determinant of portfolio investment to these countries. The authors see this as an indication of the interest of foreign investors in taking primarily controlling stakes in privatized companies. However, a number of countries which undertook large-scale privatization in 1997 did so to a significant extent by selling off companies to foreign companies. At the same time, their equity markets were still not adequately developed. In this case, foreign investors did not choose FDI because of a preference for controlling other companies, but because this was one of the few forms of acquiring equity for foreign investors.

It may be worthwhile to control in a number of the regressions for the macroeconomic performance of the respective country. It would certainly make sense if countries with low inflation, a low budget deficit, and at most a minor balance of payments problem were found to be able to attract more foreign capital than countries with severe external and internal balance problems. If this is the case, this would change several results in the paper, for example, the regressions reported in Tables A8a and A8b. In these regressions, per capita GDP is not found to be a significant determinant of FDI liabilities as a share of total liabilities and total private liabilities. It could well be that this is the case because countries with relatively good macroeconomic performance and countries with poor macroeconomic performance are not distinguished.

In a number of regressions on debt liabilities (Table A4b) and the share of FDI liabilities to total liabilities (Table A8a), the regional dummy for Africa is significant, indicating that Africa has high debt and little FDI. This could easily be misinterpreted to mean that Africa is a continent for which it is more difficult to attract foreign capital. However, other research convincingly suggests that there is no Africa effect as such. Ancharaz (1999) uses a panel of 84 countries pooled over the period 1980–1997. His main results on the determinants of FDI are reported in Table 1. The estimated equation in the first columns indicates that GDP per capita, GDP growth, government size, debt service ratio, political instability,

Table 1: Determinants of FDI (dependent variable: FDI/GDP)

Variable	(1)	(2)
GDP per capita	0.00026 (6.116)	0.00021 (1.989)
Lagged growth rate	0.0294 (3.135)	0.0323 (2.238)
Government size	–0.0534 (–3.725)	–0.044 (–2.035)
Investment ratio	0.0007 (0.069)	0.0048 (0.287)
Real exchange rate	0.00003 (0.280)	0.00003 (0.751)
Debt service ratio	–0.0081 (–2.039)	–0.0086 (–2.258)
Policy instability	–0.0556 (–1.731)	–0.0504 (–1.952)
Openness	0.0261 (6.753)	0.0244 (4.359)
African dummy		–0.199 (–0.670)
No. of observations	1142	1142
R^2	0.160	0.268
Estimation method	Fixed effects	GLS
Note: t-statistics in parentheses.		

Source: Ancharaz (1999).

and openness are significant variables when both African and non-African countries are pooled together. When a dummy variable is introduced in the equation (estimates are shown in column 2), the results are not much different from the first regression and the Africa dummy is negative but not significant. These results would therefore suggest rejecting the hypothesis that international investors are biased against Africa. Ancharaz concludes that "the marginalization of Africa in the global competition for FDI is entirely its own making, being the result of bad policies and a generally inferior investment environment." Hence, it is not the continent but rather the policies pursued by unfortunately so many countries of this continent that account for the lack of foreign capital in these economies.

At this initial step of research, Lane and Milesi-Ferretti's main finding seems to be that advanced and more open economies exhibit a higher stock of external capital. As indicated by the authors themselves, more research is needed which focuses on the driving forces of the accumulation of foreign capital (liabilities, portfolio investment and foreign investment) and the pattern of capital formation. It will be interesting to see which direction future research will take. A few suggestions are given in the next section.

4.   Some Suggestions for Further Work

It is always easier to comment critically on empirical findings than to come up with one's own findings. However, additional work in the area that Lane and Milesi-Ferretti tackle in their paper could include the following suggestions:

- Some time-series analyses would be helpful in answering questions about a structural change in the flows of capital to developing countries as suggested in the introduction to Lane and Milesi-Ferretti's paper.
- Time-series analyses would also be useful in performing tests of the type of causality between various forms of capital flows.
- In a number of cases, it would make sense to distinguish between private and official debt. The accumulation of private debt can be expected to depend on other factors than the accumulation of official debt.
- It would be desirable to control in the regressions for factors for which causality is not a problem and which seem to be important for the determination of foreign capital in a country. These factors could include
  - the existence of deposit insurance and bailout schemes for debt holders,
  - the degree of capital account openness,
  - the size of government (as too much government intervention can be detrimental to attracting foreign capital),

— indicators of macroeconomic performance,
— the level of human capital of the receiving country (which may be a pre-requisite for capital inflows), and
— the extent of infrastructure.

However, despite these suggestions for extending Lane and Milesi-Ferretti's analysis, their work can be regarded as a very useful contribution to the study of foreign capital formation in industrial and developing economies. Their data set will certainly be used by many other researchers interested in international capital flows.

Bibliography

Ancharaz, V. (1999). The Determinants of Foreign Direct Investment in a Comparative Perspective: Is There a Bias Against Africa? Mimeo, Brandeis University, Graduate School of International Economics and Finance, Waltham, Maryland.

Frederic S. Mishkin

The International Lender of Last Resort:
What Are the Issues?

F32 J16 G21
F34 O19

1. Introduction

In the 1990s, we saw financial crises erupt in emerging market countries that
have not only had devastating consequences in the countries experiencing the crises, but have roiled financial markets in other emerging market and industrialized
countries as well. The classic remedy for stopping financial crises from getting
out of control is the provision of liquidity by the so-called lender of last resort.
Indeed, in the recent crises, the international financial institutions, particularly the
IMF, have provided access to huge amounts of liquidity, on the order of $100
billion during the Asian crisis, to stem these crises.

The recent crises thus have brought the role of an international lender of last
resort into the limelight, which raises the following basic questions. First: Do we
need an international lender of last resort? Could domestic central banks instead
perform this role, as is the traditional view (Bagehot 1873)? Second: How should
the international lender of last resort operate? What are the principles that should
guide the international lender of last resort if it is to be effective? Third: Who
should be the international lender of last resort? Should the IMF be it and if so,
does it need to be reformed to perform this role effectively? Alternatively, should
another organization have the mandate to be the international lender of last resort?

This paper discusses what the issues for an international lender of last resort
are by providing some answers to these questions. But in order to understand the
role of an international lender of last resort, we must first provide an answer to a
more basic question: What is a financial crisis? Only then can we evaluate how
an international lender of last resort can be used to stop financial crises.

Remark: I thank participants at the Kiel Week conference and at a seminar at the Board of
Governors of the Federal Reserve System for their helpful comments. Any views expressed in this paper are those of the author only and not those of Columbia University or
the National Bureau of Economic Research.

292 Frederic S. Mishkin

2. What Is a Financial Crisis?

What a financial crisis is can be derived from an asymmetric information analysis of the type described in Mishkin (1991, 1996a, 1999, 2000a). A financial crisis is a nonlinear disruption of financial markets in which the asymmetric information problems of adverse selection and moral hazard become much worse, so that financial markets are unable to efficiently channel funds to those who have the most productive investment opportunities.

a. How Do Financial Crises Occur?

In most financial crises, and particularly in the Mexican and East Asian crises recently, the key factor that caused asymmetric information problems to worsen and launch a financial crisis was a deterioration in balance sheets, particularly those in the financial sector. The underlying cause of the deterioration in balance sheets was an interaction of poor government policies with a financial structure that left these emerging market countries highly vulnerable. This perspective has important implications on whether there is a need for an international lender of last resort, and, if so, what principles should guide its operation.

b. The First Stage

A key factor behind the recent financial crises was financial liberalization that resulted in the lending boom. Once restrictions were lifted on both interest-rate ceilings and the type of lending allowed, lending increased dramatically, fed by inflows of foreign capital. The problem is not that lending expands, but that it expands so rapidly that excessive risk-taking is the result, with large losses on loans in the future. There are two reasons that excessive risk-taking takes place after financial liberalization. The first is that managers of banking institutions often lack the expertise to manage risk appropriately when new lending opportunities open up after financial liberalization. In addition, with rapid growth of lending, banking institutions cannot add the necessary managerial capital (well-trained loan officers, risk-assessment systems, etc.) fast enough to enable these institutions to screen and monitor these new loans appropriately.

The second source of systemic risk that promoted excessive risk-taking was the inadequacy of the regulatory/supervisory system. Even if there is no explicit government safety net for the banking system, there clearly is an implicit safety net that creates a moral hazard problem. Depositors and foreign lenders to the banks, knowing that there are likely to be government bailouts to protect them, have little incentive to monitor banks, with the result that these institutions have an incentive to take on excessive risk by aggressively seeking out new loan business. In order to prevent this moral hazard problem, excessive risk-taking needs to be restricted by adequate government regulation. This includes the adoption of adequate accounting and legal standards, disclosure requirements, restrictions on certain holdings of assets, and capital standards. Adequate government supervision is also needed in order to monitor compliance with the regulations and assess whether the proper management controls to limit risk are in place.

Emerging market countries, particularly those in Mexico and East Asia, were notorious for weak financial regulation and supervision. When financial liberalization yielded new opportunities to take on risk, these weak regulatory/supervisory systems could not limit the moral hazard created by the government safety net and excessive risk-taking was the result. This problem was made even more severe by the rapid credit growth during the lending boom, which stretched the resources of the bank supervisors. Bank supervisory agencies were unable to add to their supervisory capital (well-trained examiners and information systems) fast enough to enable them to keep up with their increased responsibilities both because they had to monitor new activities of the banks and because these activities were expanding at a rapid pace.

Capital inflows made this problem even worse. Once financial liberalization was adopted, foreign capital fled into banks in emerging market countries because it earned high yields but was likely to be protected by the government safety net, whether provided by the government of the emerging market country or by international agencies such as the IMF. The result was that capital inflows helped fuel a lending boom which led to excessive risk-taking on the part of banks. Folkerts-Landau et al. (1995), for example, found that emerging market countries in the Asian-Pacific region with large net private capital inflows also experienced large increases in their banking sectors.

The outcome of the lending boom arising after financial liberalization was huge loan losses and a subsequent deterioration of banks' balance sheets. In the case of Mexico and the East-Asian crisis countries, the share of nonperforming loans to total loans has risen to between 15 and 35 percent (see Goldstein 1998 and Corsetti et al. 1998). The deterioration in bank balance sheets was the key fundamental that drove these emerging market countries into their financial crises.

The deterioration in bank balance sheet promotes the first stage of a financial crisis by causing banks to restrict their lending in order to improve their capital ratios. If the deterioration in bank balance sheets is severe enough, it can even lead to bank panics, in which there are multiple, simultaneous failures of banking institutions. Indeed, in the absence of a government safety net, there is some risk that contagion can spread from one bank failure to another, causing even healthy banks to fail. The source of the contagion is asymmetric information. In a panic, depositors, fearing the safety of their deposits and not knowing the quality of the banks' loan portfolios, withdraw their deposits from the banking system, causing a contraction in loans and a multiple contraction in deposits, which then causes other banks to fail. In turn, the failure of a bank means the loss of the information relationships in which that bank participated, and thus a direct loss in the amount of financial intermediation that can be done by the banking sector. The outcome is an even sharper decline in lending to facilitate productive investments, with an additional contraction resulting in economic activity.

c.   The Second Stage

In emerging market countries, the deterioration in bank balance sheets can move the crisis into its second stage, a currency crisis which then tips the economy over into a full-fledged financial crisis. A weak banking system makes it less likely that the central bank will take the steps to defend a domestic currency, which means that expected profits from selling the currency rise.[1] For example, the central bank in a country with a weakened banking system will be afraid to raise interest rates, because any rise in interest rates to keep the domestic currency from depreciating has the additional effect of weakening the banking system. Thus, when a speculative attack on the currency occurs in an emerging market country (in which speculators sell large amounts of the domestic currency for foreign currency), if the central bank raises interest rates sufficiently to defend the currency, the banking system may collapse.

The weakened state of the banking sector along with the high degree of illiquidity in Mexico and East Asian countries before the crisis then set the stage for the currency crisis. With these vulnerabilities, speculative attacks on currencies could have been triggered by a variety of factors. In the Mexican case, the attacks came in the wake of political instability in 1994, such as the assassination of political candidates and an uprising in Chiapas. Even though the Mexican central bank intervened in the foreign exchange market and raised interest rates sharply, it was unable to stem the attack and was forced to devalue the peso on December 20, 1994. In Thailand, the attacks followed unsuccessful attempts by the government to shore up the financial system, culminating in the failure of Finance One.

---

[1]   In addition, the cost of bailing out insolvent banks could produce substantial fiscal deficits which also puts downward pressure on the currency (Burnside et al. 1998).

Eventually, the inability of the central bank to defend the currency because the required measures would do too much harm to the weakened financial sector meant that the attacks could not be resisted. The outcome was therefore a collapse of the Thai baht in early July 1997. Subsequent speculative attacks on other Asian currencies led to devaluations and floats of the Philippine peso and Malaysian ringgit in mid-July, the Indonesian rupiah in mid-August, and the Korean won in October. By early 1998, the currencies of Thailand, the Philippines, Malaysia, and Korea had fallen by over 30 percent, with the Indonesian rupiah falling by over 75 percent.

Two special institutional features of credit markets in emerging market countries explain why the devaluation in the aftermath of the currency crisis helped trigger a full-fledged financial crisis. Because of past experience with high and variable inflation rates these countries have little inflation-fighting credibility and debt contracts are therefore of very short duration and are often denominated in foreign currencies. This structure of debt contracts is very different from that in most industrialized countries, which have almost all of their debt denominated in domestic currency, with much of it long-term, and it explains why there is such a different response to a devaluation in emerging market countries than there is in industrialized countries.

There are three mechanisms through which the currency crisis causes a financial crisis to occur in emerging market countries. The first involves the direct effect of currency devaluation on the balance sheets of firms. With debt contracts denominated in foreign currency, when there is a devaluation of the domestic currency, the debt burden of domestic firms increases. On the other hand, since assets are typically denominated in domestic currency, there is no simultaneous increase in the value of firms' assets. The result is that a devaluation leads to a substantial deterioration in firms' balance sheets and a decline in net worth, which, in turn, worsens the adverse selection problem because effective collateral has shrunk, thereby providing less protection to lenders. Furthermore, the decline in net worth increases moral hazard incentives for firms to take on greater risk because they have less to lose if the loans go sour. Because lenders are now subject to much higher risks of incurring losses, there is a decline in lending and hence a decline in investment and economic activity.

The damage to balance sheets from devaluation in the aftermath of the foreign exchange crisis was a major source of the contraction of the economies in East Asia, as it was in Mexico in 1995. This mechanism was particularly strong in Indonesia, which saw the value of its currency decline by 75 percent, thus increasing the rupiah value of foreign-denominated debts by a factor of four. Even a healthy firm with an initially strong balance sheet is likely to be driven into insolvency by such a shock if it has a significant amount of foreign-currency-denominated debt.

A second mechanism linking currency crises with financial crises in emerging market countries occurs because the devaluation can lead to higher inflation. Because many emerging market countries have previously experienced both high and variable inflation, their central banks are unlikely to have deep-rooted credibility as inflation fighters. Thus, a sharp depreciation of the currency after a speculative attack that leads to immediate upward pressure on prices can lead to a dramatic rise in both actual and expected inflation. Indeed Mexican inflation surged to 50 percent in 1995 after the foreign exchange crisis in 1994, and a similar phenomenon occurred in Indonesia. A rise in expected inflation after the currency crisis exacerbates the financial crisis because it leads to a sharp rise in interest rates. The interaction of the short duration of debt contracts and the interest rate rise leads to huge increases in interest payments by firms, thereby weakening firms' cash flow position and further weakening their balance sheets. Then, as we have seen, both lending and economic activity are likely to undergo a sharp decline.

A third mechanism linking the financial crisis and the currency crisis arises because the devaluation of the domestic currency can lead to further deterioration in the balance sheets of the banking sector, provoking a large-scale banking crisis. In emerging market countries, banks have many liabilities denominated in foreign currency which increase sharply in value when a depreciation occurs. On the other hand, the problems of firms and households mean that they are unable to pay off their debts, also resulting in loan losses on the assets side of the banks' balance sheets. The result is that banks' balance sheets are squeezed from both the assets and liabilities side and the net worth of banks therefore declines. An additional problem for the banks is that many of their foreign-currency denominated debt is very short-term, so that the sharp increase in the value of this debt leads to liquidity problems for the banks because this debt needs to be paid back quickly. The result of the further deterioration in bank balance sheets and their weakened capital base is that they cut back lending. In the extreme case, the deterioration of bank balance sheets leads to a banking crisis that forces many banks to close their doors, thereby directly limiting the ability of the banking sector to make loans; the affect on the economy is even more severe.

The bottom line of this asymmetric information analysis is that the recent financial crises were the result of a systemic collapse in both financial and nonfinancial firm balance sheets that made asymmetric information problems worse. The result was that financial markets were no longer able to channel funds to those with productive investment opportunities, which then had led to a severe economic contraction.

3.   Do We Need an International Lender of Last Resort?

We have seen that a seizing up of information in the financial system when a financial crisis occurs leads to disastrous consequences for the economy. To recover, the financial system needs to be restarted so that it can resume its job of channeling funds to those with productive investment opportunities. The asymmetric information view thus provides a rationale for government intervention to get the financial system back on its feet, thereby preventing systemic risk episodes from spinning out of control.

In industrialized countries, domestic central banks have the ability to do this with a lender-of-last-resort operation in which the central bank lends freely during a financial crisis. I will argue however, that central banks in emerging market countries are much less likely to have this capability. Thus there is a strong argument that an international lender of last resort may play a crucial role in limiting the damage from financial crises in these countries. However, even if there is a need for an international lender of last resort, an international lender of last resort does create a serious moral hazard problem that can make financial crises more likely. Thus an international lender of last resort which does not sufficiently limit these moral hazard problems can actually make the situation worse, a subject that is discussed in the section following this one.

a.   The Lender of Last Resort in Industrialized Countries

In the face of a possible banking panic, a central bank in an industrialized country can provide liquidity to banks to prop them up, giving the government time to close them down in an orderly fashion. Indeed, this is exactly what the Federal Reserve did in 1984 when it lent $5 billion to Continental Illinois, one of the ten largest banks in the United States at the time, giving the FDIC time to take it over, thereby making sure that depositors did not suffer any losses. The result was that no further runs on other banks occurred and a panic was avoided.[2]

---

2   There were problems in the way the Continental Illinois bailout was handled because not only depositors, but also all creditors, were prevented from taking any losses. In addition, shortly afterward, the Comptroller of the Currency testified that not only Continental Illinois, but eleven other banks were considered "too big to fail." This led to a well-known moral hazard problem because creditors of large banks now had less incentive to monitor the bank and pull out their funds if the bank was taking on excessive risk. These problems, however, do not imply that the lender-of-last-resort operation was not the appropriate response. Instead, they suggest moral hazard needs to be limited when a government safety net is provided for the banking system.

Not only can a central bank in an industrialized country be a lender of last resort to banks, but it can also play the same role for the financial system as a whole. Two prominent examples described in Mishkin (1991) occurred when the Federal Reserve provided liquidity in the aftermath of the Penn Central bankruptcy in June 1970 and the stock market crash of 1987. When Penn Central, a large commercial paper issuer, went broke, the commercial paper market seized up, making it difficult for many corporations to roll over their commercial paper, raising the potential for cascading bankruptcies. The Fed came to the rescue by providing liquidity to banks and encouraging them to lend to corporations that could not roll over their commercial paper.

After the stock market crash on October 19, 1987, securities firms needed to extend massive amounts of credit on behalf of their customers for their margin calls in order to keep the stock market and the related index futures market functioning in an orderly fashion. However, understandably enough, banks were growing very nervous about the financial health of securities firms and so were reluctant to lend to the securities industry at a time when it was most needed. There was thus a major danger of a spreading collapse of securities firms and a further market meltdown. To prevent this, Alan Greenspan, the new Fed Chairman, announced before the market opened on Tuesday, October 20, the Federal Reserve's "readiness to serve as a source of liquidity to support the economic and financial system." In addition, the Fed made it clear that it would provide liquidity to any bank that would make loans to the securities industry, although this did not prove to be necessary. Indeed, what is striking about this episode is that the extremely quick intervention of the Fed resulted not only in a negligible impact on the economy of the stock market crash, but also meant that the amount of liquidity that the Fed needed to supply to the economy was not very large (see Mishkin 1991).

A key reason that central banks in industrialized countries can successfully engage in lender-of-last-resort operations is that the institutional structure of financial systems in most industrialized countries has the following two features: (1) debt contracts are almost solely denominated in domestic currency, and (2) because inflation has tended to be moderate, many debt contracts are of fairly long duration. As a result, the monetary expansion resulting from injecting liquidity into the financial system helps stimulate recovery of the economy even further. Injecting reserves, either through open market operations or by lending to the banking sector, causes the money supply to increase, which in turns leads to a higher price level. Given that debt contracts are denominated in domestic currency and many debt contracts are of fairly long duration, the reflation of the economy causes the debt burden of households and firms to fall, thereby increasing their net worth. As outlined earlier, higher net worth then leads to reduced adverse selection and moral hazard problems in financial markets, undoing the increase in adverse selection and moral hazard problems induced by the financial crisis. In addition, injecting liquidity into the economy raises asset prices such as land and stock market values, which also causes an improvement in net worth and a reduction in adverse selection and moral hazard problems. Also, as discussed in Mishkin (1996b), expansionary monetary policy promotes economic recovery through other mechanisms involving the stock market and the foreign exchange market.[3] Thus the expansionary monetary policy associated with a lender-of-last-resort operation works in the right direction and helps promote recovery from the financial crisis.

b.   The Lender of Last Resort in Emerging Market Countries

However, institutional features of the financial systems in emerging market countries imply that it may be far more difficult for the central bank to promote recovery from a financial crisis with a lender-of-last-resort operation. As mentioned before, many emerging market countries have much of their debt denominated in foreign currency. Furthermore, their past record of high and variable inflation has resulted in debt contracts of very short duration and expansionary monetary policy is likely to cause expected inflation to rise dramatically.

As a result of these institutional features, in emerging market countries a lender-of-last-resort operation is far less likely to be successful. Given the past record on inflation, in an emerging market country central bank lending to the financial system in the wake of a financial crisis which expands domestic credit might arouse fears of inflation spiraling out of control.

In this case the expansionary monetary policy is likely to cause the domestic currency to depreciate sharply. As we have seen before, the depreciation of the domestic currency leads to a deterioration in firms' and banks' balance sheets be-

---

[3]   Note that not all industrialized countries are alike in their ability to use expansionary monetary policy to recover from a financial crisis. If a country has a commitment to peg its exchange rate to a foreign currency, then expansionary monetary policy may not be an available tool to promote recovery because pursuing such a policy might force a devaluation of its currency. This problem is of course particularly acute for a small country in a pegged exchange rate regime. Even if a country has a flexible exchange rate, expansionary monetary policy to promote recovery might cause a depreciation of the domestic currency which is considered to be intolerable by the authorities, particularly in smaller countries. Clearly, a large reserve currency country like the United States has the most flexibility to use expansionary monetary policy to reflate the economy as a tool to recover from or reduce the probability of a financial crisis.

cause much of their debt is denominated in foreign currency, thus raising the burden of indebtedness and lowering banks' and firms' net worth. In addition, the upward jump in expected inflation is likely to cause interest rates to rise because lenders need to be protected from the loss of purchasing power when they lend. As we have also seen, the resulting rise in interest rates causes interest payments to soar and the cash flow of households and firms to decline. Again the result is a deterioration in households' and firms' balance sheets, and potentially greater loan losses to banking institutions. Also, because debt contracts are of very short duration, the rise in the price level from expansionary monetary policy does not affect the value of households' and firms' debts appreciably, so this mechanism provides little benefit to their balance sheets, unlike the case in industrialized countries.

The net result of the expansionary monetary policy associated with the lender-of-last-resort operation in the emerging market countries with the above institutional structure is that it hurts the balance sheets of households, firms, and banks. Thus, the lender-of-last-resort operation has the opposite result of that found in industrialized countries during a financial crisis: it causes a deterioration in balance sheets and therefore amplifies adverse selection and moral hazard problems in financial markets caused by a financial crisis, rather than ameliorating them as in the industrialized country case.

c.   The Case for an International Lender of Last Resort

The above arguments suggest that central banks in emerging market countries have only a very limited ability to extricate their countries from a financial crisis. Indeed, a speedy recovery from a financial crisis in an emerging market country is likely to require foreign assistance because liquidity provided from foreign sources does not lead to any of the undesirable consequences that result from the provision of liquidity by domestic authorities. Foreign assistance does not lead to increased inflation which through the cash-flow mechanism hurts domestic balance sheets, and it helps to stabilize the value of the domestic currency, which strengthens domestic balance sheets.

Thus, since a lender of last resort for emerging market countries is needed at times and it cannot be provided domestically but must be provided by foreigners, there is a strong rationale for having an international lender of last resort. A further rationale for an international lender of last resort exists if there is contagion from one emerging market country to another during a financial crisis. Although the jury is still out on this one, it does appear that a successful speculative attack on one emerging market country does lead to speculative attacks on other emerging market countries, which can lead to collapses of additional currencies. Thus, currency crises do have the potential to snowball, and because these currency crises lead to full-fledged financial crises in emerging market countries, the risk of contagion is indeed a serious one. An international lender of last resort has the ability to stop contagion by providing international reserves to emerging market countries threatened by speculative attacks so that they can keep their currencies from plummeting. This assistance can thus keep currency and therefore financial crises from spreading.

4.   How Should the International Lender of Last Resort Operate?

Although the asymmetric information analysis above provides a rationale for an international lender of last resort, there is a cost. The existence of an international lender of last resort creates a serious moral hazard problem because depositors and other creditors of banking institutions expect that they will be protected if a crisis occurs. In the recent crisis episodes, governments in the crisis countries have used support from international financial institutions to protect depositors and other creditors of banking institutions from losses. This safety net creates a well-known moral hazard problem because the depositors and other creditors have less incentive to monitor these banking institutions and withdraw their deposits if the institutions are taking on too much risk. The result is that these institutions are encouraged to take on excessive risks which make financial crises more likely.

Thus, to limit the moral hazard problem created by an international lender of last resort and to help it cope with financial crises more effectively, our analysis suggests eight principles to guide the operation of the international lender of last resort: (1) restore confidence to the financial system, (2) provide liquidity to restart the financial system, (3) provide liquidity as fast as possible, (4) restore balance sheets, (5) punish owners of insolvent institutions, (6) encourage adequate prudential supervision, (7) engage in lender-of-last-resort operations only for countries that are serious about implementing necessary reforms, and (8) engage in lender-of-last-resort operations infrequently and only for short periods of time.

a.   Principle 1: Restore Confidence to the Financial System

The most important principle for an international lender of last resort is that it restore confidence in the financial system. Without confidence, participants will pull out of the financial markets, and the financial system will not be able to channel funds to those with productive investment opportunities. Restoring confidence is thus essential to keeping the financial system operating efficiently, which is the key to preventing or promoting recovery from a financial crisis. However, restoring confidence is easier said than done and so requires several actions on the part of the international lender of last resort.

b.   Principle 2: Provide Liquidity to Restart the Financial System

The first step in the process of keeping the financial system operating and restoring confidence is to provide ample liquidity so that markets stay in operation and restart. However, injecting liquidity into the financial system may not be effective unless the following three principles are followed.

c.   Principle 3: Provide Liquidity as Fast as Possible

An important historical feature of successful lender-of-last-resort operations is that the faster the lending is done, the lower the amount is that actually has to be lent. This fact provides support for the second principle that the faster liquidity is provided in an international lender-of-last-resort operation, the better. An excellent example occurred in the aftermath of the stock market crash on October 19, 1987. What is remarkable about this episode is that the Fed did not need to lend directly to the banks to encourage them to lend to the securities firms who needed funds to clear their customers' accounts. Because the Fed acted so quickly, within a day, and reassured banks that the financial system would not seize up, banks knew that lending to securities firms would be profitable, and so it was in their interest to make these loans immediately even without borrowing from the Fed, which they then did. With confidence restored and the fear of crisis diminished, the actual amount of liquidity that the Fed needed to inject into the banking system through open market operations was quite small and this liquidity was removed shortly after the crisis was over.

In contrast, during the recent financial crises in Mexico and East Asia, the amount of liquidity made available was very large, exceeding $100 billion. In these episodes, putting together the rescue packages took time, on the order of several months. By this time, the crises had gotten much worse, with the result that much larger sums were needed to shore up the system. As the example of the Federal Reserve's lender-of-last-resort operation in October 1987 indicates, a much smaller amount of funds would have done the trick in the recent crises if these funds could have been disbursed faster. The need for quick provision of liquidity to keep the amount of funds manageable suggests that credit facilities at an international lender of last resort must be designed to provide funds quickly.

d.   Principle 4: Restore Balance Sheets

The asymmetric information analysis of financial crises indicates that the resolution of and recovery from a financial crisis requires a restoration of the balance sheets of both financial and nonfinancial firms. Restoration of balance sheets of nonfinancial firms requires a well-functioning bankruptcy law that enables the balance sheets of these firms to be cleaned up so they can regain access to the credit markets. Restoration of balance sheets of financial firms requires that insolvent financial institutions be closed down and that public funds be injected so that healthy institutions can buy up the assets of insolvent institutions. Successful resolution of the crisis also requires the creation of entities like the Resolution Trust Corporation in the United States, which can sell off assets of failed institutions and get them off the books of the banking sector.

Crucial to a successful resolution of a financial crisis is that half-measures to clean up balance sheets are avoided. With balance sheets only partially restored, weak financial institutions may hold back on their lending because they may need to increase capital ratios in the future. Alternatively, leaving weak financial institutions in operation may encourage them to take on excessive risk because they have little to lose, thus diminishing confidence in the future health of the financial system. Insolvent financial institutions must be put out of their misery.

The international lender of last resort and potentially other international organizations can help in this process by sharing their expertise and by encouraging the governments in crisis countries to take steps to create a better legal structure and a better resolution process for failed financial institutions.

e.   Principle 5: Punish Owners of Insolvent Institutions

There are two reasons why encouraging punishment of owners of insolvent financial institutions is important for an international lender of last resort. In any emerging market countries, owners of insolvent financial institutions are frequently provided with funds which enables them to keep their institutions in operation or to pocket substantial wealth which they often ship out of the country before the institution fails. Bailing out the owners in this way leads to a tremendous moral hazard problem because knowing that this will occur, these owners have incentives to take on huge risks because they have so little to lose. The result is that a financial crisis is far more likely. This moral hazard can be limited by ensuring that owners of insolvent institutions are punished, that is, that they are not allowed to keep their institutions operating and that when they are closed down, they suffer substantial losses. Of course, encouraging losses for large creditors of financial institutions goes even further in reducing moral hazard incentives for risk-taking at financial institutions because these creditors now have incentives to monitor the institutions and pull out their funds if they are taking on excessive risk.

In addition, encouraging punishment of owners of insolvent financial institutions is necessary in order to generate sufficient public support for sufficient funds to clean up the balance sheets of the financial sector. For example, in Japan, this was not done with the bailout of the jusen in the mid-1990s. Public outrage that owners of these institutions, which include many large banks and reputedly some criminal figures, got off scot free is one reason why the public in Japan has been unwilling to support injection of sufficient public funds into the banking system to get it fully back on its feet. The consequences of a continuing weak banking system have been disastrous for Japan and are an important cause of Japan's economic stagnation (see, e.g., Mishkin 1998). In the United States, in contrast, owners of insolvent savings and loans did incur substantial losses and sometimes were even thrown in jail. Such actions helped provide political support for the full bailout of the savings and loan industry in 1989.

f.   Principle 6: Encourage Adequate Prudential Supervision

The moral hazard problem created by the existence of a safety net for financial institutions can also be limited by the usual elements of a well-functioning prudential regulatory/supervisory system (see, e.g., Mishkin 2000b): adequate disclosure requirements, adequate capital standards, prompt corrective action, careful monitoring of the institution's risk management procedures, and monitoring of financial institutions to enforce compliance with the regulations. However, there are often strong political forces in emerging market countries which resist putting these kinds of measures into place. This has also been a problem in industrialized countries—for example, an important factor in the U.S. savings and loan debacle was political pressure to weaken regulation and supervision (see, e.g., Kane 1989)—but the problem is far worse in many emerging market countries. What we have seen in the Asian crisis countries is that the political will to adequately regulate and supervise financial institutions has been especially weak because politicians and their family members are often the actual owners of financial institutions. An international lender of last resort is particularly well suited to encourage adoption of the above measures to limit moral hazard because it has so much leverage over the emerging market countries to whom it lends or who might want to borrow from it in the future.

There are two reasons why an international lender of last resort will be more successful if it actively encourages adoption of the above prudential regulatory/supervisory measures. The first is that its lender-of-last-resort actions provide governments with the resources to bail out their financial sectors. Thus an international lender of last resort strengthens the safety net, which increases the moral hazard incentives for financial institutions in emerging market countries to take on excessive risk. It can counter these incentives by strengthening the regulatory/supervisory apparatus in these countries. The second is that the presence of an international lender of last resort may create a moral hazard problem for governments in emerging market countries who, because they know that their financial sectors are likely to be bailed out, have less incentive to take steps to prevent domestic financial institutions from taking on excessive risk. The international lender of last resort can improve incentives to reduce excessive risk-taking by making it clear that it will only extend liquidity to governments that put the proper measures in place to prevent excessive risk-taking (see Goldstein 1998). Only with this kind of pressure can the moral hazard problem arising from lender-of-last-resort operations be contained.

g.   Principle 7: Engage in Lender-of-Last-Resort Operations Only for Countries That Are Serious about Implementing Necessary Reforms

One problem that arises for an international lender of last resort is that it knows that if it doesn't come to the rescue, the emerging market country will suffer extreme hardship and possible political instability. Politicians in the crisis country may exploit these concerns and engage in a game of chicken with the international lender of last resort: they resist necessary reforms, hoping that the international lender of last resort will cave in. Elements of this game were present in the Mexico crisis of 1995 and this was also an important feature of the negotiations between the IMF and Indonesia during its recent crisis.

An international lender of last resort will produce better outcomes if it makes it clear that it will not play this game. Just as giving in to your children may be the easy way out in the short run, but leads to children who are poorly brought up in the long run, so the international lender of last resort will be more successful by not giving in to short-run humanitarian concerns and let emerging market countries escape from necessary reforms. An international lender of last resort will improve its performance by being willing to walk away from a country that is not willing to help itself. Indeed, if it caves to one country during a financial crisis, politicians in other countries will see that they can get away with not implementing the needed reforms, making it even harder for the international lender of last resort to limit moral hazard.

h.   Principle 8: Engage in Lender-of-Last-Resort Operations Infrequently and Only for Short Periods of Time

Because there is a tradeoff between the benefits of a lender-of-last-resort role in preventing financial crises and the moral hazard that it creates, a lender-of-last-resort role is best implemented only when it is absolutely necessary. An international lender of last resort thus has strong reasons to resist calls on it to provide funds under normal conditions. Furthermore, once a crisis is over, the liquidity that has been injected into the financial system needs to be removed in order to ensure that financial markets do not become dependent upon it. In other words, the lender-of-last-resort role will be more successful if it is implemented only very infrequently and for short periods of time.

5.   Who Should Be the International Lender of Last Resort?

Given that there is a need for an international lender of last resort, what institution would be best suited to perform this role and follow the principles outlined above?

Traditionally, central banks have acted as lenders of last resort because they have the advantage of being able to create the necessary liquidity at will. In addition, they have had experience in successfully performing this role. These facts would argue for the creation of a world central bank to act as an international lender of last resort. However, because it is highly unlikely that the major countries of the world would be willing to give up control of monetary policy to an international organization for the foreseeable future, creation of a world central bank is unrealistic.

Reasoning that the international financial institution that closest resembles a central bank is the Bank for International Settlements (BIS), which has the function of being a bank for central banks and thus might be characterized as the central banks' central bank, some economists (e.g., Shadow Open Market Committee 1998) have advocated that the BIS take on the international lender-of-last-resort role. However, even though the BIS does hold substantial amounts of deposits from central banks of industrialized countries, it is not quite right to see the BIS as engaging in traditional central banking functions. Rather than seeing the BIS as an organization that can inject liquidity into the world financial system, it is better thought of as a club for central bankers where they can share information and coordinate their activities. Indeed, the BIS role in the international financial system has become more important over time for exactly this reason: for example, it has become a useful site for coordination of prudential regulations, as with the Basel Committee for Bank Supervision, which meets under the auspices of the BIS.

The last three principles in the previous section indicate that an international lender of last resort needs a lot of information in order to decide whether engaging in a lender-of-last-resort operation is absolutely necessary and whether a country is serious about adequate prudential supervision. However, consistent with its role as a central bankers' club, the BIS is a small organization with under 500 employees, a small fraction of the employment of the central banks that are members. Thus, the BIS does not currently have the capability, nor is it seeking the capability, to act as an international lender of last resort.

Indeed, the only international organization that currently has the staff to acquire the necessary information to effectively perform the international lender-of-last-resort role is the International Monetary Fund (IMF). This is why, by default, it has ended up engaging in this role during the recent crisis episodes. One objection to the IMF's performing a lender-of-last-resort role is that it cannot create unlimited liquidity as can a central bank. But, as persuasively argued by Fischer (1999), it is not absolutely necessary that an international lender of last resort have unlimited resources to create liquidity, just that it has enough to do the job. Indeed, Fischer (1999) points out that under the gold standard, central banks in reality did not have an unlimited capability to create liquidity and yet were able to perform the lender-of-last-resort role, and so the situation for the IMF in this regard is not all that different. It is true that the IMF's resources might limit its capabilities to manage a crisis, but it is not clear that this has been a problem in the recent crisis episodes. Furthermore, if the IMF requires more resources to adequately deal with financial crises, Fund quotas could be raised to get access to these resources.

a.   Is the IMF Set Up Well to Do It?

If the IMF is the only crap game in town, i.e., the only international financial institution capable of performing the international lender-of-last-resort role, is it well designed to do this effectively? The principles in the previous section outlining how an international lender of last resort should operate suggest that the answer is no. Not surprisingly, the desire to improve the performance of the IMF in managing recent crises has led to numerous proposals for reform of the institution.

Although the IMF was originally set up to provide liquidity only in the short-term to cope with balance of payments imbalances, after the Latin American debt crisis in 1982 the IMF began to broaden its policy agenda and engage in longer-term lending to poor countries. The broadening of the IMF's agenda has contin-ued over the years. With the replacement in 1999 of the Enhanced Structural Adjustment Facility (ESAF), under which long-term lending under preferred rates to poor countries was conducted, by the Poverty Reduction and Growth Facility (PRGF), the Fund now has an explicit goal of reducing poverty.[4] The Fund has also begun to venture into labor and environmental issues. Another feature of the IMF's lending is that it is frequent. Seventy countries have received credit under IMF programs for 20 or more years (see Vasquez 1999).

The expansion of the IMF's activities is clearly at odds with the basic princi-ple that operation of an international lender of last resort works best if the lender-of-last-resort role is performed infrequently and for only short periods of time. Long-term lending is obviously inconsistent with lending for short periods of time, while continuous lending to countries is inconsistent with the need for lender-of-last-resort operations to be infrequent. The habit of frequent, or con-tinuous, lending makes it more likely that the IMF will engage in "crisis" lending when it might not be absolutely necessary, thereby increasing the moral hazard problem. In addition, continuous lending may make it hard for the IMF to resist providing loans when a country's government is not sufficiently committed to necessary reforms of the financial sector, thus making it more difficult for the IMF to adhere to Principle 7.

The mindset created by the IMF's engagement in frequent, long-term lending makes it highly problematic for other reasons. To limit the moral hazard prob-lems inherent in long-term lending, conditionality on a wide range of issues is necessary to make sure the funds are used for the appropriate purposes. Design-ing this conditionality takes time and, as Principle 3 indicates, the success of a lender-of-last-resort operation requires provision of liquidity as fast as possible.

The IMF is aware of the problem that time delays arising from putting condi-tionality into place creates and so established a Conditional Credit Line (CCL) facility in 1999, which allows preapproved countries with good policies to re-ceive credit immediately without having to negotiate new conditions for the loan.

---

[4]   See IMF (1999a) for a description of the mandate in Poverty Reduction and Growth Facility. For a discussion of the evolution of the IMF's activities over time, see Boughton (1998), Krueger (1998), Vasquez (1999), Overseas Development Council (2000), and International Financial Institution Advisory Commission (2000).

However, so far no country has applied to this facility. Countries seem to be concerned that applying to this credit facility will create concerns that the country is vulnerable to a financial crisis. In addition, they may be concerned that being dropped from this facility will be a sign to the markets that they are in serious trouble, possibly precipitating the financial crisis that they are trying to avoid. Indeed, this problem may make it very hard for the IMF to take away a country's approval for the CCL once it has been given to them, potentially giving the country less incentive to keep good policies in place. The CCL has thus not yet solved the problem of developing a credit facility which enables them to perform the lender-of-last-resort role quickly and yet limit moral hazard.

Wide-ranging conditionality also means that there is a lack of focus in IMF programs. As the analysis of financial crises indicates, the fundamental driving factor behind the recent financial crises has been microeconomic problems in the financial sector, and this is why an international lender of last resort must focus on encouraging necessary reforms to ensure adequate prudential supervision of the financial system as in Principle 6. It is true that recent IMF programs for the crisis countries have conditions dealing with financial sector reforms, but with so many other IMF conditions, politicians in the crisis countries may pick and choose the conditions they want to follow. They are then likely to drag their feet on conditions involving financial sector reform which harm their close friends, and even family, and this is exactly what transpired in Indonesia with its IMF program after its financial crisis. Furthermore, because wide-ranging conditionality frequently imposes conditions that no industrialized country would tolerate, governments in emerging market countries can raise the flag of interference with sovereignty to garner political support to avoid the necessary reforms of the financial sector.

Another problem with wide-ranging conditionality has been raised by Feldstein (1998). Because these conditions are often considered onerous, their possible imposition may discourage countries from coming to the IMF at an early stage of the financial crisis. Thus, it becomes more likely that IMF lending will be slow in coming, the opposite of what is proposed in Principle 3, thus making the financial crisis worse and requiring even larger provision of funds by the IMF.

Another problem with the expansion of IMF activities is that the IMF is less capable of acquiring the information and knowledge base it needs to perform the lender-of-last-resort role effectively. The IMF has been criticized for not understanding the difference between the recent crises in Mexico and East Asia and earlier balance of payments crises (see, e.g., Radelet and Sachs 1998 and Furman and Stiglitz 1998). This led them to violate Principle 1, i.e., the need to restore confidence in the financial system, when they advocated the closure of sixteen banks in Indonesia without setting up a safety net for the rest of the banking system. This led to a banking panic and a further collapse of the Indonesian rupiah, which through the mechanisms described in the asymmetric information analysis in the first section, exacerbated the financial crisis. Indeed, one finding of a committee that I chaired that evaluated the IMF's research activities (IMF 2000) was that the research staff was so overstretched that the organization did not have time to do the necessary thinking to get a better understanding of the nature of the crises they were facing.

The above problems with the expansion of the IMF's activities over time has led to calls for reform to narrow the IMF's focus. What is striking about several recent proposals for reform from many different sources, including the U.S. Treasury (Summers 1999), the Council on Foreign Relations (1999), the Overseas Development Council (2000), and the International Financial Institutions Advisory Commission (the so-called Meltzer Commission) (2000) set up by the U.S. Congress, is that all of them call for a narrowing of the IMF's focus on issues involved with crisis management. Furthermore, the U.S. Treasury, the Overseas Development Council, and the Meltzer Commission have all recommended that the IMF get out of the long-term lending business. Although these reforms appear to have wide-spread support, it is not clear that the IMF is conducive to them. The report of the external evaluation committee for IMF surveillance (IMF 1999b) recommended narrowing the focus on surveillance activities, yet this was questioned by the IMF's Executive Board and the staff (see IMF 1999b).

6.   Concluding Remarks

This paper has argued that an international lender of last resort can play an important role in improving the functioning of the international financial system. It has also outlined eight principles to help ensure that the international lender of last resort will be both effective and limit the moral hazard that its presence creates. At the present time, the International Monetary Fund appears to be the only international financial institution capable of acting as the international lender of last resort, but its current activities interfere with effective performance of this role. Reforming the IMF to enhance its performance in managing financial crises is therefore likely to be at the top of the agenda for reform of international financial architecture.

Bibliography

Bagehot, W. (1873). *Lombard Street: A Description of the Money Market.* London: H.S. King.

Boughton, J. (1998). From Suez to Tequila: The IMF as Crisis Managers. Mimeo, International Monetary Fund, Washington, D.C.

Burnside, C., M. Eichenbaum, and S. Rebelo (1998). Prospective Deficits and the Asian Currency Crisis. Working Paper 98–5. Federal Reserve Bank of Chicago. September 1998.

Corsetti, G., P. Pesenti, and N. Roubini (1998). What Caused the Asian Currency and Financial Crisis? Part I and II. NBER Working Papers 6833 and 6844, Cambridge, Mass.

Council on Foreign Relations (1999). *Safeguarding Prosperity in a Global Financial System: The Future International Financial Architecture.* Report of an Independent Task Force Sponsored by the Council on Foreign Relations. Washington, D.C.: Institute for International Economics.

Feldstein, M. (1998). Refocusing the IMF. *Foreign Affairs* (March/April): 20–33.

Fischer, S. (1999). On the Need for an International Lender of Last Resort. *Journal of Economic Perspectives* 13(4): 85–104.

Folkerts-Landau, D., G.J. Schinasi, M. Cassard, V.K. Ng, C.M. Reinhart, and M.G. Spencer (1995). Effect of Capital Flows on the Domestic Financial Sectors in APEC Developing Countries. In M.S. Khan and C.M. Reinhart (eds.), *Capital Flows in the APEC Region.* Washington, D.C.: International Monetary Fund.

Furman, J., and J.E. Stiglitz (1998). Economic Crises: Evidence and Insights from East Asia. *Brookings Papers on Economic Activity* (2):1–114.

Goldstein, M. (1998). *The Asian Financial Crisis.* Washington, D.C.: Institute for International Economics.

International Financial Institutions Advisory Commission (2000). *Report.* Washington, D.C.

IMF (International Monetary Fund) (1999a). *The Poverty Reduction and Growth Facility (PRGF)—Operational Issues.* Washington, D.C.: International Monetary Fund.

—— (1999b). *External Evaluation of IMF Surveillance: Report by a Group of Independent Experts.* Washington, D.C.: International Monetary Fund.

—— (2000). *External Evaluation of IMF Research Activities: Report by a Group of Independent Experts.* Washington, D.C.: International Monetary Fund.

Kane, E.J. (1989). *The S&L Insurance Mess: How Did It Happen?* Washington, D.C.: Urban Institute Press.

Krueger, A. (1998). Whither the World Bank and the IMF? *Journal of Economic Literature* 36(4): 1983–2020.

Mishkin, F.S. (1991). Asymmetric Information and Financial Crises: A Historical Perspective. In R.G. Hubbard (ed.), *Financial Markets and Financial Crises.* Chicago: University of Chicago Press.

—— (1996a). Understanding Financial Crises: A Developing Country Perspective. In M. Bruno and B. Pleskovic (eds.), *Annual World Bank Conference on Development Economics 1996.* Washington, D.C.: World Bank.

—— (1996b). The Channels of Monetary Transmission: Lessons for Monetary Policy. *Banque De France Bulletin Digest* 27: 33–44.

—— (1997). The Causes and Propagation of Financial Instability: Lessons for Policy-makers. In C. Hakkio (ed.), *Maintaining Financial Stability in a Global Economy.* Kansas City: Federal Reserve Bank of Kansas City.

—— (1998). Promoting Japanese Recovery. In K. Ishigaki and H. Hino (eds.), *Towards the Restoration of Sound Banking Systems in Japan—The Global Implications.* Kobe: Kobe University Press.

—— (1999). Lessons from the Asian Crisis. *Journal of International Money and Finance* 18(4):709–723.

—— (2000a). The Korean Financial Crisis: An Asymmetric Information Perspective. *Emerging Markets Review* 1: 21–52.

—— (2000b). Financial Market Reform. In A. Krueger (ed.), *Economic Policy Reform: What We Know and What We Need to Know.* Chicago: University of Chicago Press.

Radelet, S., and J.D. Sachs (1998). The East Asian Financial Crisis: Diagnosis, Remedies and Prospects. *Brookings Papers on Economic Activity* (1): 1–74.

Overseas Development Council (2000). *The Future Role of the IMF in Development: An ODC Task Force Report.* Baltimore: Johns Hopkins University Press.

Shadow Open Market Committee (1998). Policy Statements and Position Papers. March 15–16, 1998. Bradley Policy Research Center Working Paper PPS 98–01, University of Rochester.

Summers, L. (1999). The Right Kind of IMF for a Stable Global Financial System. Speech presented at the London School of Business, December 14.

Vasquez, I. (1999). The International Monetary Fund: Challenges and Contradictions. Paper presented to the International Financial Advisory Commission, September 28.

p 291:

Comment on Frederic S. Mishkin 016 019 G21 F32 F34

Catherine L. Mann

1. The Lender of Last Resort: National and International

Rick Mishkin addresses the question of what the issues are for the *international* lender of last resort. He notes that to answer this question requires thinking first about the national lender of last resort. I completely agree. One way to systematically relate the international lender of last resort (LLR) to the national LLR is to look at the *occasions* under which an LLR might take action, the *instruments* an LLR might use, and the *supporting structures* necessary to reduce the risk of systemic financial crisis without excessively raising the risk of individual moral hazard (Mann 1998, 1999).

This systematic pairing requires analyzing what is different in the international context from the national context. Is the key difference the inability of a national LLR to perform its functions in responding to domestic financial distress, so that a supranational LLR must step in and perform the duties of a national LLR? Is there even more to the international environment that expands the scope of what an international LLR might need to do because of the potential for international systemic failures that could arise even if national LLRs are able to perform their functions?

Following the review of the differences between national and international LLRs, the question is, does the IMF have the instruments and supporting structures necessary to respond to the occasions when it might be called upon? In the end, I agree that the IMF is the only supranational institution that can serve as the international lender of last resort. But Rick and I differ somewhat on why the IMF should play this role, how it fulfills it, what the consequences are of the approach taken in the recent set of financial crises, and what the next steps should be.

I argue that the reason why the IMF must play this role is fundamentally one of international systemic distress. The IMF is the only organization that can coordinate action when sovereign nations are involved, and when fast moving global financial crises demand large and immediate injections of credit to the international financial system. The source of growth in an increasingly global world is increasingly through international financial intermediation—the collapse of the international financial system cannot be risked.

However, the IMF operates without key supporting mechanisms of a constant supervisory presence and a fiscal redistributive authority that are integral to the environment of the national LLR and which mitigate moral hazard. Consequently, IMF intervention in the current international financial environment has magnified moral hazard.

So, even as we bolster the IMF's credit line for when needed, we must seek ways to limit the occasions we resort to it. Recommendations to improve transparency and disclosure and to strengthen national financial systems are necessary but will not eliminate the need for an international LRR, just as these do not eliminate the need for a national LRR.

The next phase requires market-oriented solutions. The private financial market has the technical ability to create financial instruments that will mitigate moral hazard and diversify risk to participants who can best bear it. The issue is how to induce the private sector to do its job.

2.   The Lender of Last Resort in the National Context

A national LLR exists because distress at a single financial institution could spill over to other financial institutions, thus impairing the conduct of the whole system. Although financial institutions should accurately price and manage their own risks, no financial institution prices into its services the possibility that it will have to absorb the consequences of a spillover from another institution. Nor should it, since doing so would lead to an inappropriately high price and thus too small an amount of intermediation. Because the social value of the financial system as a whole exceeds the value created by individual firms, there is a rationale for centralized scrutiny and possibly intervention.

A national LLR has several instruments to limit contagion (limited deposit insurance, reserves management, sufficient equity) and others to intervene (liquidity, moral or balance-sheet support). These instruments come with some unintended consequences, the most familiar of which is moral hazard.

Supervision and regulation is the key supporting structure necessary to balance moral hazard with the social value of the financial system. Because financial institutions might respond to actual or implied central bank support by following riskier business strategies, supervision and regulation of their activities is usually required. What exactly is the scope of this supervision and regulation? First, it is ongoing; supervision of a distressed institution comes too late. Second, it demands increasingly active intervention into institutions that are foundering (so-called prompt corrective action); precipitous closure is too much too late. Third, it requires a balance between prohibition of activities and management of those activities; financial intermediation involves transforming risks (interest rate, currency, credit). Appropriate supervision and regulation involves risk management, not risk avoidance.

Finally, the distribution of loss or gain that results from intervention is increasingly important. The incidence of the loss (and implied gain) across various segments of the population may be a relevant issue when a national LLR considers intervening. For a national financial system and national LLR, the incidence of loss and gain is likely to be mostly among domestic residents. Thus, the domestic fiscal authority could use its instrument to redistribute the loss and gain to maintain support for the necessary actions of the LLR. However, as the share of foreign depositors increases or the importance of foreign financial intermediaries increases, the incidence of impact of LLR support (national or international) can become much more complex, and cross-border.

3.   The Lender of Last Resort in the International Context

The biggest difference in the international context is that financial intermediaries offer not only maturity transformation, but also transformation of currency exposures. As international trade in goods and services continues to grow faster than world output, the demand for these financial services increases and the efficiencies they offer and economic benefit they create increase as well. Avoiding the international financial system is not the right answer.

However, what happens if a national LLR needs to support a particular institution to contain a domestic financial crisis but does not have access to the currencies of the open obligations? "Swap" networks between central banks can serve this purpose, but the amounts are limited. Mishkin suggests that this is a key place where the IMF should intervene to provide the "right" currency mix, in particular because the activities of the national LLR without "credibility" could lead to inflation and a collapse of financial balance sheets. However, whereas the IMF has access to many currencies through its callable capital, most currencies have rather little value in an international crisis. Who decides how much of a war chest and of which currencies the IMF should have on hand?

A second difference in the international context is that the transparency of balance sheets is worse, which makes it more difficult for the LLR to decide when and how to intervene. Often, financial institutions that intermediate internationally operate on the fringes of the national supervisory and regulatory environment or even arbitrage across different regimes, finding the cracks between them. In response, central banks, working through the BIS, have developed the Basle Accord on Capital Adequacy and Core Principles, which sets guidelines for transparency, provisioning, and equity capital for institutions.

Third, it is clear that international financial systems are prone to contagion beyond the domestic experience. Failures in one national market can affect other national markets through contagion of financial institutions as well as through changes in sentiment in capital markets. A coordinated response of national LLRs could obviate the need for an international LLR. But because of differences in behavior and mandate, national authorities may be unable to agree on an appropriate response within the short time frame necessary to prevent disastrous contagion. Consequently, an international LLR is needed beyond the national LLR.

Finally, a problem in the international context is that there is no international fiscal authority to finance the costs of intervention nor to ameliorate imbalances in the incidence of costs and benefits of intervention. IMF lending in support of financial sector adjustment programs does allow governments to shift their national burden through time, but the burden is retained within a country. When some private sector financial intermediaries agree to rollover and reschedule debt, it may alter the incidence of loss and place more of the burden on the overseas private investor. To do more and in a less ad hoc way, an international fiscal authority as complement to the international LLR would be needed but is unlikely to be created.

4.   A Key Missing Piece—the International Supervisor: Can the IMF Be It?

The Basle Core Principles represent an effort to "internationalize" supervision and regulation, but it falls to the national supervisors to enforce these rules (Goldstein 1997). In particular, since financial crises are part of a "repeated game," supervisors only have credibility if they can enforce the rules day after day, not just in times of crisis. Thus, any international authority, or even a lead regulator, would have to operate with the national LLR and its regulatory presence and structure, or else a country would have to foreswear its own regulator and give those responsibilities to an international one.

No supranational supervisor can demand compliance and transparency without the ability to deliver liquidity when needed; that is the quid pro quo of supervision. However, IMF credit has been provided too slowly in crisis times and to governments, not to financial intermediaries. And, IMF conditionality goes well beyond "prompt corrective action" or closure of individual financial institutions.

While there may be a rationale for such conditionality, it mixes the LLR and supervisory functions with broader development and macrostability functions. As a result, no country subjects itself to broad conditionality without broad credit arrangements, and all countries step down from IMF conditionality as soon as possible. Consequently, an international supervisory agency housed at the IMF would lack credibility and force.

Mishkin, among others (Meltzer 2000, Hills et al. 2000), argues that this is the rationale for constraining the functionality of the IMF to financial sector development and supervision. But, this limit on the IMF's functionality would not necessarily yield the ongoing financial supervision of individual institutions and could limit its ability to coordinate credit when needed.

Because of the fast-moving nature of these crises, the need for immediate and large injections of credit, and because sovereign nations are involved, the IMF is the only institution that can offer coordinated action, but it cannot effectively enforce its conditionality, in particular with regards to supervisory issues and restructuring of distressed financial systems. So, the IMF has performed the credit side of the international LLR, but lacks the key supporting structure of supervision.

Moreover, the incidence of gains and losses involved in international financial crises have spread beyond national boundaries in important ways. There is no quasi-fiscal authority that can redistribute it to, perhaps, the investors and institutions who "should" bear the losses. Consequently, the situation that the IMF is responding to and the environment in which it operates magnify the moral hazard problems that are present with the national LLR.

5.   The Private Sector Must Play a Greater Role

It is unrealistic to suppose either that there will be no international lending of last resort or to suppose that the supranational authorities with national responsibility that are needed as a counterweight to moral hazard will be created. Countries and their companies will be supported in times of financial distress because of the possible international systemic consequences. So, the issue is how to complement the IMF's role as coordinator of credit with greater control over moral hazard.

The fundamental way to avoid systemic financial distress is to make sure that borrowers and lenders accurately price and manage risks in their own portfolios. What makes financial markets work well? Market participants with different tastes for risk, armed with full information, and offered "complete" markets of financial instruments. All three ingredients have been missing in recent years.

Financial insurance instruments, such as credit risk insurance or restructuring insurance, could help stabilize the international financial system (see also Eichengreen 1999). First, these instruments differentiate the market participants. Not all borrowers would offer insurance; not all lenders would buy it. But for those that did, when the financial crisis hits, the insured lenders would not abandon the insured borrowers, at least not for the duration of the policy. Insurance could significantly alter the herd mentality in the market by diversifying the exposure of market participants and by moving risk from those who fear it to those who manage it. The change in the pace of the race to the exits could dramatically alter the self-fulfilling nature of some financial crises.

Second, financial insurance splits the pricing of financial products into pieces which can then be priced separately. A financial relationship needs to consider two situations: the borrower-lender relationship during normal times and the relationship when the borrower is distressed. The interest rate on a loan or bond only prices the relationship during normal times. There is no interest rate high enough to pay off the principal of the loan or bond in the case where a borrower defaults. In contrast, financial insurance prices the cost of default.

Who might develop these financial products? Private financial institutions have the technical ability to create financial instruments that will price default risk and diversify it to participants who can bear it, as they have done in the context of international currency exposure.

But, a key aspect of the demand for financial insurance is the presence or absence of international bailouts or orderly (e.g., IMF or public-sector-coordinated) workout agreements. To the extent that creditors are made whole or partially whole through nonmarket mechanisms, the demand for insurance instruments will not develop. Why should any creditor pay for insurance if they get an international bailout for "free" (as in Tesobonos) or if the costs of renegotiation of the terms of repayment are coordinated by an official third party (as in Korea in early 1998)?

Can these instruments be created ex ante the next financial crisis? Before such a crisis, volatility in the foreign exchange market would create the incentive for currency swaps and options so that investors could pay for insurance, in advance, against unexpected movements in exchange rates. Similarly, the more difficult, drawn-out, ad hoc, and therefore costly the financial-disaster workouts are, the greater the incentives are for investors to demand and institutions to offer instruments ex ante that will help generate a market-oriented solution to the workout process. Therefore, rather than intervening more frequently, official institutions must stand aside more often.

Bibliography

Eichengreen, B. (1999). *Toward a New International Financial Architecture: A Practical Post-Asia Agenda.* Washington, D.C.: Institute for International Economics.

Goldstein, M. (1997). *The Case for an International Banking Standard.* Policy Analyses in International Economics no. 47. Washington, D.C.: Institute for International Economics.

Hills, C., P.G. Peterson, and M. Goldstein (2000). Safeguarding Prosperity in a Global Financial Systm: The Future International Financial Architecture. Report of an Independent Task Force Sponsored by the Council on Foreign Relations. Washington, D.C.: Institute for International Economics.

Mann, C.L. (1998). Guest Spotlight. *Emerging Markets Debt Monthly,* Merrill Lynch, April 23:48–54.

—— (1999). Market Mechanisms to Reduce the Need for IMF Bailouts. Policy Brief 99-1. Institute for International Economics, Washington, D.C.

Meltzer, A. (2000). Report of the International Financial Institution Advisory Commission. Submitted to the U.S. Congress and U.S. Department of Treasury, March 8.

LIST OF CONTRIBUTORS

MARCO BECHT
European Centre for Advanced Research in Economics and Statistics, Free University of Brussels

CLAUDIA M. BUCH
Financial Markets Research Area, Kiel Institute of World Economics, Kiel

WILLIAM R. CLINE
Deputy Managing Director and Chief Economist, Institute of International Finance, Washington, D.C.

PAUL DE GRAUWE
Professor, Centre for Economic Studies, University of Leuven, Belgium

GUNTER DUFEY
Professor, University of Michigan School of Business, Ann Arbor, Michigan

SEBASTIAN EDWARDS
Professor, Anderson Graduate School of Management at UCLA, Los Angeles, California

THEO S. EICHER
Professor, Department of Economics, University of Washington, Seattle, Washington

MICHAEL FRENKEL
Professor, Wissenschaftliche Hochschule für Unternehmensführung, WHU Koblenz, Vallendar, Germany

LESLIE HULL
Professor, School of Economics and Finance, Victoria University of Wellington, New Zealand

PHILIP R. LANE
Professor, Department of Economics, Trinity College, Dublin, and Centre for Economic Policy Research, London

CATHERINE L. MANN
Institute for International Economics, Washington, D.C.

COLIN MAYER
Peter Moores Professor of Management Studies (Finance), Said Business School, University of Oxford, Oxford

GIAN MARIA MILESI-FERRETTI
Research Department, International Monetary Fund, Washington, D.C., and Centre for Economic Policy Research, London

FREDERIC S. MISHKIN
Alfred Lerner Professor of Banking and Financial Institutions, Graduate School of Business, Columbia University, New York, and Research Associate, National Bureau of Economic Research

TOMMASO PADOA-SCHIOPPA
Member of the Executive Board of the European Central Bank, Frankfurt am Main, and Honorary professorship at the University of Frankfurt am Main

CHRISTIAN PIERDZIOCH
Financial Markets Research Area, Kiel Institute of World Economics, Kiel

JEAN PISANI-FERRY
Senior Economic Adviser, Ministry of Economy, Finance, and Industry, Paris

MAGDALENA POLAN
Centre for Economic Studies, University of Leuven, Belgium

HELMUT REISEN
Head of Research Division, OECD Development Centre, Paris

LINDA L. TESAR
Professor, Department of Economics, University of Michigan, Ann Arbor, Michigan

INGO WALTER
Professor, Director of New York University Salomon Center, New York

INDEX

KIELER STUDIEN · KIEL STUDIES

Kiel Institute of World Economics

Editor: *Horst Siebert* · Managing Editor: *Harmen Lehment*

More information on publications by the Kiel Institute at http://www. uni-kiel.de/ifw/pub/ pub.htm, more information on the Kiel Institute at http://www.uni-kiel.de/ifw

---

Tübingen: Mohr Siebeck (http://www.mohr.de)
Berlin · Heidelberg: Springer-Verlag (http://www.springer.de)

Kiel Institute of World Economics

Symposia and Conference Proceedings

Horst Siebert, Editor

Monetary Policy in an Integrated World Economy
Tübingen 1996. 280 pages. Hardcover

Towards a New Global Framework for High-Technology Competition
Tübingen 1997. 223 pages. Hardcover.

Quo Vadis Europe?
Tübingen 1997. 343 pages. Hardcover.

Structural Change and Labor Market Flexibility
Experience in Selected OECD Economies
Tübingen 1997. 292 pages. Hardcover.

Redesigning Social Security
Tübingen 1998. 387 pages. Hardcover.

Globalization and Labor
Tübingen 1999. 320 pages. Hardcover.

The Economics of International Environmental Problems
Tübingen 2000. 274 pages. Hardcover.

The World's New Financial Landscape: Challenges for Economic Policy
Berlin · Heidelberg 2001. 324 pages. Hardcover.

Tübingen: Mohr Siebeck (http://www.mohr.de)
Berlin · Heidelberg: Springer-Verlag (http://www.springer.de)